ISBN 978-1-331-94356-3
PIBN 10257776

English
Français
Deutsche
Italiano
Español
Português

www.forgottenbooks.com

Mythology Photography **Fiction**
Fishing Christianity **Art** Cooking
Essays Buddhism Freemasonry
Medicine **Biology** Music **Ancient
Egypt** Evolution Carpentry Physics
Dance Geology **Mathematics** Fitness
Shakespeare **Folklore** Yoga Marketing
Confidence Immortality Biographies
Poetry **Psychology** Witchcraft
Electronics Chemistry History **Law**
Accounting **Philosophy** Anthropology
Alchemy Drama Quantum Mechanics
Atheism Sexual Health **Ancient History**
Entrepreneurship Languages Sport
Paleontology Needlework Islam
Metaphysics Investment Archaeology
Parenting Statistics Criminology
Motivational

FROM AN EASY CHAIR

BY

SIR RAY LANKESTER, K.C.B., F.R.S.

" The world is so full of a number of things,
I am sure we should all be as happy as kings."

R. L. STEVENSON

LONDON

ARCHIBALD CONSTABLE & CO., LTD.

1909

RICHARD CLAY AND SONS, LIMITED
BREAD STREET HILL, E.C., AND
BUNGAY, SUFFOLK.

Published October, 1908.
Reprinted January, 1909.

PREFACE

THIS little book is a reproduction, with some emendations, of articles which appeared in the *Daily Telegraph* in the six months between the beginning of last October and the end of April. If it should meet with success, further collections of the same kind will be published from time to time.

E. R. L.

August, 1908.

CONTENTS

		PAGE
1.	Science and the Study of Nature	1
2.	The Desire to Know the World of Nature	3
3.	Scares and Wonders	5
4.	Work at the Pasteur Institute	9
5.	The Sea Serpent	10
6.	Giraffes and the Okapi	11
7.	The Great Geologists of Last Century	14
8.	Experiments with Precious Stones	19
9.	Diamonds	23
10.	Science and Fisheries	27
11.	Discoveries as to Malaria	29
12.	Malta Fever	34
13.	A Cure for Sleeping Sickness	36
14.	Tsetse-Flies and Disease	38
15.	Monkeys and Fleas	41
16.	The Jigger Flea	42
17.	Public Estimate of the Value of Science	43
18.	The Common House-fly and Others	45
19.	Cerebral Inhibition	48
20.	Colour-photography and Photographs of Mars	49
21.	Origin of Names by Errors in Copying	50
22.	False News as to Extinct Monsters	51
23.	Mistletoe and Holly	52

CONTENTS

PAGE

24. The Cattle Show 55
25. The Experimental Method 59
26. Hypnotism and an Experiment on the Influence or the Magnet 60
27. Luminous Owls and Other Luminous Animals and Plants 65
28. Reminiscences of Lord Kelvin . . . 68
29. The So-called Jargon of Science . . . 70
30. Rats and the Plague 73
31. Ancient Temples and Astronomy . . . 78
32. Alchemists of To-day and Yesterday . . . 84
33. A Story of Sham Diamonds and Pearls . . 88
34. The Nature of Pearls 89
35. A King Who was a Zoologist 93
36. The Transmission to Offspring of Acquired Qualities 97
37. Variation and Selection Among Living Things . 103
38. The Movement, Growth, and Dwindling of Glaciers 108
39. Votes for Women 117
40. Tobacco and the History of Smoking . . . 124
41. Cruelty, Pain, and Knowledge 131

FROM AN EASY CHAIR

1. *Science and the Study of Nature*

THIS volume consists of brief notes in plain language on a variety of scientific matters. I speak of new discoveries, real or so-called by mistake; of old well-established facts and explanations of strange occurrences which are more familiar to men of science than to people who have not had the time and opportunity to ascertain what is, and what is not proved and known about Nature and her ways. I do not address my reader from the professor's chair, but from an easy chair. Just as in the club or my friend's smoking-room, I might talk of these things, so do I propose to talk here. My hope is that what I have to say will interest those who are not experts in science, and yet have a desire for trustworthy information and opinion on the vast variety of topics which come up day by day for consideration and discussion, and can only be explained or rightly understood by the aid of that systematised knowledge which is called science.

———

Science and the scientific point of view have a very wide, indeed, an unlimited range. Though the making of discoveries of real importance and the full understanding of the steps by which they are made involves, as a rule,

B

long study and special training, yet there is a vast deal of healthy excitement and pleasure connected with the progress of science, in which all can share by receiving, as it were, messages from the front. By contributing true records and observations of fact which serve, in however small a way, as ammunition and material of war for the use of the fighting line, we can all help and take part in the advance of science.

A great feature of what is called science is that it is true. The actual result achieved by science is the record of " that which is "—it can be examined, tested, and proved. But science does not merely collect accurate records of fact. In order to discover new things, new relations, and hidden causes she has to make use of guesses and flights of imagination. The " hypotheses " or guesses are not wild ones, but reasonable suppositions based on careful consideration of existing knowledge. They are never mistaken by trained workers in science for " facts," nor put forward as such. On the contrary, they are tested and so confirmed or rejected by experiment or trial. Hence the necessity of accuracy in observation for the purposes of science ; hence the proverbial " scientific accuracy." It is of no use to form a guess based upon erroneous statements. It is mere waste of time to accept and build theories upon loose wonder-mongers' gossip. And, further, the evidence which you obtain in order to confirm or dismiss your " guess " must be equally beyond suspicion as to its accuracy. It must be an observation of fact free from prejudice and illusion.

Your guess, if proved to be true, adds to the solid record of science new facts and new proofs of relationships, which again lead on the imagination of men of science to new guesses, and so to new confirmation or rejection, and to the growth of the vast record of accurate knowledge. To seek out in the endless whirling complexity of things which surround us in earth, sky, and sea, the truth, the knowledge of " that which is," of the relation of these things to one another as cause and effect and their action and influence on ourselves—this is the aim of science. To substitute real understanding and the power of control of the surrounding world for the misleading and cruelly harmful conceptions existing in the minds of simple unskilled mankind—this is the daily achievement of science.

2. *The Desire to Know the World of Nature*

The practical value of science in securing the happiness of human communities is not, however, the reason which operates most strongly in exciting men and women to give themselves to the cultivation and improvement of this or that branch of it. A rich banker one day was looking round the Natural History Museum with me. It was his first visit. After a time he said, "It's very fine! wonderful! But what's it all for? Where does the money come in? That's what I can't understand. Why does the Government spend money on this if it don't lead to making money?" I tried to convince him that there exists in us all a divine " curiosity," a desire to know regardless of profit or loss, a thirst which we may cultivate and satisfy, in the full assurance that whilst its satisfaction is a delight in itself, we are all the while fulfilling the destiny of man, helping in the conquest of Nature. My friend had apparently lost that instinctive thirst which is the primary impulse to the pursuit of science, that capacity for pleasure which Robert Louis Stevenson truly notes in the words of the child of his " Garland of Verse":

> " The world is so full of a number of things,
> I am sure we should all be as happy as kings!"

The existence of that little child and of numberless

"grown-ups" who have become or have never ceased to be, in this matter, even as he, is the reason why science has its helpers and workers of all ranks, and it is of them that I chiefly think in writing these notes.

At a dinner of the Savage Club a year or so ago my friend Dr. Nansen, the Norwegian Minister, quoted some lines from a Scandinavian poet, which he translated somewhat as follows: "As you journey through life do not go too fast, do not press on blindly; there are so many beautiful things by the way. Turn your head, stay a few minutes. Leave the dusty road. Take in and enjoy the wonders and delights which are at your feet." Motorists, please take note!

For those who can enter more thoroughly into the pursuit of science there are even greater joys. To the very few there is the privilege not merely of realising well-established truths, and of perhaps assisting in securing their foundations or extending their application, but of discovering vast unexplored regions, new possibilities, new revelations of the unfathomed depths of Nature's workings. Though few can hope to be leaders in these enthralling adventures, yet we can be close to those who are, and, holding their hands, sympathise with their soul's vision.

> "Then felt I like some watcher from the skies,
> Or the stout Cortes, when, with eagle eyes,
> He stared at the Pacific . . . .
> Silent, upon a peak in Darien."

Such a one need have none of the conventional setting of romantic enterprise. He may be standing before a much-stained table, covered with bottles, in an atmosphere of acrid fumes, with a test-tube in his hand, or he may be just raising his head with a far-off gaze, as he sits, bent o'er a microscope, in London.

3. Scares and Wonders

There are certain subjects which come within my ken upon which paragraphs are published in the papers nearly every other day of a wildly romantic and misleading character. These subjects may be classified as: (1) Living and extinct monsters. (2) Cures for cancer and tubercle. (3) Unsuspected dangers of infection by disease-germs. It would hardly be pleasant for me to quote these paragraphs in order to deny their statements. They are often headed, " For the Little Ones," or " From a Foreign Correspondent." The old-established and better title for such announcements is " For the Marines." I shall endeavour to mention as they occur to me, among other things, new and duly-certified facts relating to monsters, and to the investigation of disease. With reference to reports which have been seriously put forward during the past year, I may say that the alleged discovery of a mammoth in North America 71ft. long and 40ft. tall is nonsense. In the announcement to which I allude, the measurements have been altered from some original and more correct statement and made to appear astonishing by error or design.

No new facts of importance bearing upon the treatment of either cancer or tubercle have been lately discovered which can be explained to the general public. Work is proceeding nevertheless. No new source of danger from disease-germs has been detected since this time last year. It is true that the dust in railway carriages, and especially in sleeping-cars, which are not properly cleaned every day after occupation by travellers, is full of microbes, and, like the dust of rooms which have been crowded by human beings, may be a source of disease infection. The remedy for this is careful cleansing after each journey, and a special construction of the cars like a tiled bath-room, so as to avoid the accumulation of dirt. At present this is, and long has been, neglected.

Another serious and more recent danger is that arising from the crowding of passengers in underground railway tubes. Both in Paris and London this has been recognised as a real and pressing danger. Trouble has been given by the dust raised in the Paris Tube, but the danger caused by dust has been avoided in London. It is a definitely-ascertained fact that many bacteria, including disease-producing kinds, are rapidly killed by exposure to strong sunlight. Hence underground tubes and the chinks and recesses of railway carriages are more liable to harbour disease-germs than the open-air roadways and the carriages which ply on them. Great cleanliness and the use of germicide washing fluids are the obvious precautions to be taken in the absence of sunlight.

As to mammoths and elephants—the former is a misspelling of the word "mammont," the name given by the natives of Northern Siberia to the extinct elephant, hairy, but otherwise closely similar to the Indian elephant, which within the period of prehistoric man (50,000 to 150,000 years) was abundant over the whole of the northern part of the Northern Hemisphere. Mammoths' tusks (ivory) are still largely imported from Siberia. The biggest African elephant may, perhaps, stand 13ft. at the shoulder. No mammoth or other extinct elephant seems to have exceeded this. The stuffed African elephant in Cromwell road measures 11ft. 2in. at the shoulder. Mr. Carnegie's great extinct reptile Diplodocus is only 12ft. 9in. from the ground at the highest part of its back. The biggest tusk of a recent elephant ever seen was bought by me for the Natural History Museum seven years ago. It weighs 228lb., and measures 10ft. 2in. along the curve. It was recognised three years ago by Mr. Jephson (one of

Stanley's companions) as one of a pair which he had weighed in Central Africa. It was in the possession of Emin Pasha when that unfortunate gentleman was "rescued" by Stanley and Jephson. After the subsequent assassination of Emin, his ivory treasure found its way to Zanzibar, and this tusk being part of it, was sold and brought to London.

A real new monster of great size is the carnivorous reptile described by Professor Osborne, of New York, as Tyrannosaurus. There is no mistake or exaggeration about this report. The specimen is in the New York Museum, and has been described in detail and drawn to scale by Professor Osborne. The skeleton stands up like that of a huge bird or a kangaroo on the two hind legs —as does that of the vegetarian reptile Iguanodon. The Iguanodon and the Tyrannosaurus are of about the same height, namely 17ft. But the new monster has enormous tiger-like teeth, twelve on each side of the jaw, above and below, and the jaws are three feet long, whilst the whole head is broad and short. Iguanodon, on the other hand, has been long known from English and Belgian rocks, and can be seen in Cromwell Road. It has a beak like a tortoise, and the small teeth of a vegetable-feeder. Both animals had very short front limbs or arms, but in Tyrannosaurus these are really ridiculously out of proportion, according to more familiar standards, for the whole arm is not bigger than one of the toes of the hind foot. This new giant carnivorous reptile is found in rocks of the same age as our green-sand and chalk in Wyoming, U.S.A. It preyed upon huge vegetable-eating reptiles, the remains of which are found in the same strata, and have been reconstructed.

The mere size of these extinct reptiles is a very natural cause of wonder and admiration. At the same time, it is well to remember that the body of the largest African elephant is as big, or very nearly as big, as the body of the biggest of these extinct reptiles. Some of these giant extinct reptiles had very long tails and necks, which the elephant cannot boast. No extinct animal is known which approaches in bulk the great whales of various kinds at present inhabiting the sea. The striking thing about many huge extinct animals is that they are represented to-day by similarly constructed animals of much smaller size. Thus we know giant extinct sloths, which contrast strangely with the small living sloths of to-day, giant extinct rat-like animals and giant extinct kangaroos far exceeding the bulk of living rats and kangaroos. But it is distinctly not true that all recent animals are degenerate and small as compared with extinct related kinds. The modern horse is far larger than its extinct ancestors, which we can trace back in a gradual diminishing series to a little beast no bigger than a spaniel. So, too, the earliest elephants known are quite small creatures.

The interesting point about extinct animals is really not so much that they were often large of their kind, but that they are often of kinds quite unknown at the present day among living animals. On the other hand sometimes (but by no means always) they can be shown to be connected as ancestors to living animals by a series of intermediate forms. The remains of the connecting forms are found embedded in successive rock-strata, intermediate in age between the present day and the remote period when the earliest members of the series were alive and flourishing—and we can follow out in many instances (for example, in the pedigree of the horse, and again of the elephant) the gradual but very extensive changes by which the descendants of a long extinct kind of animal have been " transformed " into modern recent animals, familiar to us.

4. *Work at the Pasteur Institute*

Professor Elias Metschnikoff was busy, when I saw him at the Institut Pasteur in Paris last September, with an experimental investigation of " appendicitis." He finds that chimpanzees can exhibit this disease, and he is led by experiments on those animals to believe that a gas-producing micro-organism—the bacillus aërogenicus—already known as occurring in the human intestine—is especially active in exciting the disease. Parasitic worms or other foreign bodies must first wound the delicate lining of the appendix before the virulent gas-forming bacillus can penetrate and start inflammation and abscess. Metschnikoff was also investigating a disease of tropical regions, known as " the Yaws." Most people would imagine that this name refers to a disease like the gapes, but it is quite different, being an ulceration of the skin caused by a spirillum.

Spirilla—corkscrew-like threads of excessive minuteness—are parasitic organisms, like bacteria, bacilli, and micrococci. They are of different kinds—some harmless, some deadly. One is common in the mouth of the healthiest of us—another causes one of our most terrible diseases. They can be distinguished by the microscope, though much alike. What microscopists call " dark-ground illumination "—that is, illumination by horizontal rays of light, obtained by a prism attached below the glass slip on which the object is placed for examination with the microscope, has been found at the Institut Pasteur to be a very ready way of showing the spirilla in fresh blood or sputum. The spirilla are alive, and are seen when highly magnified, shooting rapidly across the field of view with a corkscrew action, like brilliant silver threads. The detection of the microbe which causes an infective disease, is often the first step to the control of the disease, or to knowledge which enables man to avoid the disease altogether. Some striking examples of this have occurred of late years.

5. *The Sea Serpent*

The sea-serpent rarely puts in an appearance now, though a Cornish "manifestation" was reported last year. A recent account of a strange marine monster, declared by some to be, of course, the sea-serpent, seen but to disappear, was that given by Lord Crawford's companions two years ago. In that case, and in others in which a huge fin-like structure, supported by fin-rays, has been seen projecting from the mysterious animal, it is not improbable that what was seen was a large seal of the "eared" kind, raising one of its long, webbed hind-feet from the water, a trick which some of them are known to have. Other reputed sea-serpents have been, in reality, a school of porpoises, or a line-like flight of sea-birds, or a mass of seaweed, or a whale in association with one or other of these—or, again, a real marine snake 5ft. long (such are well known and very poisonous), or a ribbon-fish 12ft. long. There is "no reason why there should not be" a huge and seldom-seen kind of animal living in the sea—like a serpent in appearance. No one can say, as the result of observation, that there is not, since no one has thoroughly explored the dark, unfathomed depths of ocean. Yet we gain very little when we have admitted our ignorance, and agreed that there is no reason why something should not be. The real question is, "Does the thing in question exist?" not "Could it possibly exist?" Does the great sea-serpent exist? The answer to that is, There is not much evidence to show that it does. Most persons who have looked into the matter would be willing to bet 100,000 to 1 against its being captured, dead or alive, and brought before the Royal Society within ten years' time. Unless it be so captured and "tabled" it matters very little whether it exists or not. It must be "discovered" in order to become really interesting.

6. *Giraffes and the Okapi*

The baby giraffe at the gardens in the Regent's Park is a most interesting and beautiful creature. In that respect she only resembles on a small scale her grown-up relatives. Next to elephants, giraffes take precedence for strangeness, beauty, and imposing size. Certainly they have done so with me ever since I turned one Sunday afternoon long ago from the great novelty of the day, the first hippopotamus sent from Egypt, round whom the world of fashion was crowding, and gazed into the beautiful eyes that hung over me, supported by a gracefully-curving neck. My tender regard for the beautiful creature was not shaken even when I felt a sudden jerk to the elastic band passing under my chin and saw my new Leghorn straw hat, with its ornamental bunch of Egyptian wheat and broad pink ribbon, disappear between the lips of the beauty. A slow right and left movement of the jaw followed, accompanied by a tranquil kindly look suggestive of a desire for more. That was one of the old stock of Regent's Park giraffes, who bred freely at the gardens and made money for the society. They died out thirty years ago or more. From time to time since then there have been one or two mis-shapen giraffes in London, but they did not eat children's hats nor produce young of their own. A new dynasty of Kordofan giraffes has now arrived, and a better spirit prevails.

The most interesting thing about the giraffe is the okapi. The remark sounds absurd, but it is true. The okapi is the new animal from the Congo forest of

Central Africa, discovered in 1901 by Sir Harry Johnston. It is as big as a very large stag, has a neck like a deer, and is striped on the haunches and legs, not spotted as is the giraffe. Yet its teeth and its horns prove it to be a close ally, not of deer, but of the giraffe. Any points of agreement between giraffes and the okapi are, therefore, important. I have examined the baby giraffe at the Zoo, and find that she has stripe-like bands of hair on the face and on other parts of the head. Both her father and mother are from Kordofan, and have some six or seven strongly-marked bands of dark hair over the eyes and on the muzzle. It is important to note any colour-striping in the giraffe's skin, since the giraffe's colour-markings are mostly in the form of great spots, whilst the okapi is only marked by stripes or bands something like those of a zebra, but confined to the haunches and the legs, the rest of the body being dark brown. The tendency to develop colour stripes in the giraffe is important, since it shows us that the stripes do not separate the okapi absolutely from the camelopard; they are a common possession or possibility of the two animals. It was my examination of a half-brother of the little giraffe now alive at the Gardens which led to the discovery of striping on the head and face of giraffes. The mother in that case had died before the birth of her young one, and the dead calf was given to me by the secretary of the Zoological Society. Sixty-eight years ago Sir Richard (then Professor) Owen received a new-born giraffe from the Gardens, and reported on it to the Zoological Society. No one had examined one since that date; none were obtainable from the Zoo, and I could get none from African travellers and sportsmen, in spite of urgent requests. I was accordingly greatly pleased to secure one from the London Gardens. A great peculiarity of the young giraffe is that it is born with a pair of well-grown horns, nearly an inch long, and covered with coarse black hair. No other horn-bearing mammal—no antelope, buffalo, ox, sheep, goat, stag, or other deer—is born with horns, so far as we know, and we know a good many of these animals well. Before birth the young giraffe's horns are flat from back to front, and quite soft and flexible. They can be pressed backwards, so as to be made to lie flat on the head. Directly after birth a hard, bony deposit commences inside the horn, and after some years' growth it becomes firmly fused to the skull. But the hard bony core never breaks through the hairy skin which covers it. The bony core of the okapi's pair of horns, on the contrary, does "cut" or break through the skin, exposing a sharp, hard point, a quarter of an inch in length. In the deer tribe, as everyone knows, the point of the bony horn-core spreads out as a large, branching growth from which all covering is shed, and forms the "antler." The deer tribe shed the antlers every year from [the top of the horn-core, and grow a new and larger pair to take the place of the old ones. Moreover, in them the horn-core itself is a stem-like upgrowth of the bone of the skull (of the frontal bone). In the okapi and the giraffe the horn-core is a separate bone, free at first and fusing with the skull only when the adult condition is reached. The little antlers or bare-points of the okapi's horn-cones or cores seem to be shed in segments as growth goes on, and are only minute things compared with the antlers of stags. The giraffe's horns, on the other hand, always remain covered by skin and hair and have a broad, rounded top, not a sharp point.

The real clinching feature in the okapi and giraffe which decides at once their close affinity to one another is found in the outer tooth on each side of the group of eight teeth placed in the front of the lower jaw. In both this particular tooth has a broad, chisel-like crown, divided into two portions by a deep vertical slit. None of the other ungulate or hoofed animals have this very curious shape of tooth. It is a sort of family "mark" or "feature" in okapis and giraffes, as may be seen in specimens shown in the gallery of the Natural History Museum, where we have now no less than three fine, well-stuffed okapis and several varieties of giraffe.

7. The Great Geologists of Last Century

The centenary of the foundation of the Geological Society of London, celebrated last year, was a genuine festival in the scientific world. Though geology had its teachers and searchers before 1807 (Hutton and Werner, and the Neptunian and Plutonic schools, with their theories as to the origin of rocks on the one hand by marine deposit, or on the other by igneous agency, flourished before that date), yet it is true that the adequate conception of the problems of geology and the proper use of accurate observations and of judicious theory based on those observations, in relation to the problems of geology, coincided with the foundation of the society. It was not the first "special" scientific society founded in London; there was already the Linnean Society (founded in 1788) for the cultivation of zoology and botany. Yet it incurred the displeasure of the worthy president of the Royal Society, Sir Joseph Banks, who at first joined it, and then withdrew from it, when, in 1809, it ceased to be a dining-club, meeting at a London tavern, and acquired rooms of its own at No. 4, Garden-court, Temple. Apparently there was a notion in those days that the "Royal Society for the promotion of Natural Knowledge," founded in 1662, should exercise a sort of paternal control over any society formed for the special promotion of one branch of science.   Independence has, however, been found to be the healthiest condition, and we now have not only the Linnean and the Geological, but the Zoological, the Chemical, and the Physical Societies, vigorous and important corporations, publishing their "Transactions," and meeting for discussion.   There is, it is true, a danger that the Royal Society may be left eventually, owing to these independent establishments, in the sole possession and control of the doctors and the engineers. It is a curious fact that the word " physiology," which in Cicero's time (he says " Physiologia naturæ ratio ") and in the Middle Ages meant what we now call " natural history," has been abandoned by other sciences, and appropriated by the medical men.   In England, but not abroad, the doctors have even usurped the words "physician" and "physic."   In France, on the contrary, and more correctly, Lord Rayleigh and Sir William Crooks are called distinguished "physicians," and the theory of the luminiferous ether is " physic."

The Geological Society issued its first volume of Transactions in 1811.   The origin of the society is there stated to be due to " the desire of its founders to communicate to each other the results of their observations, and to examine how far the opinions maintained by the writers on geology are in conformity with the facts presented by nature."   A more exact and intelligible statement of the attitude of scientific men, then and now, could not be formulated.

There are few, if any, among us now who knew many of the original members of the Geological Society, but I remember meeting, when I was a youth, Leonard Horner, the first secretary of the society, and father-in-aw of Sir Charles Lyell.   I also knew Dr. Peter Mark Roget, an original member, who was the oldest fellow of the Royal Society when he died in 1869.   Sir Henry Holland, the father of the present Lord Knutsford, became a member in 1809, and published a paper on the rock-salt district in the first volume. He was an eminent medical man, and a great traveller. He wrote, amongst other things, upon the turquoise mines of Persia and upon longevity. He was a friend of my father's, and I had the advantage of talking the latter subject over with him before I wrote a little book on "Comparative Longevity" in 1869.

It was not until 1825 that the Geological Society obtained a charter, and was incorporated. Two great names appear in the first council of the newly-incorporated society—Murchison and Lyell. Murchison became the Director of the Geological Survey, and as "Sir Roderick" was a familiar and picturesque figure in the scientific world of the second and third quarters of last century. He wore an Inverness cape and a tall hat with a large and much-curled brim, an old-fashioned stock, and a tail-coat. In his hand he always grasped a large, handsome cane, with which he expressed his applause during the discussions at the society, or emphasised his own remarks. He was fond of alluding to himself as "an old soldier of the hammer," and almost always entered into a discussion with these words, "It is now, sir, a quarter of a century since, in company with my illustrious friend, Sir Somebody Something, I had the privilege and pleasure of showing that"—whatever it might be. Discussions at the Geological in the sixties and seventies were real, animated, almost violent discussions. I need hardly say that they were perfectly delightful. Godwin Austen was a fine, incisive speaker, who seemed ready to back his statements and views with his fists, if need be. Lyell, the greatest of all, was most modest, and almost timid in pressing an opinion, but full of personal experience and minute knowledge of facts. John Phillips, the nephew of the father of English geology, William Smith, was mellifluous and persuasive; Jukes, robust and defiant; Huxley (secretary and then president).

clear, trenchant, and uncompromising. I remember an occasion when Sir Roderick, with tears in his voice, if not in his eyes, declared he would not stay in the room to hear that fossil fishes were discovered in his own special domain—the Silurian rocks, where he had long since shown that they did not occur—and he left the meeting. Many Silurian fishes have now been found, but we all loved Sir Roderick for the heart and feeling which he threw into his work and his public utterances.

The aim of geology is to describe accurately the long succession of changes in the crust of "this cooling cinder," the earth, and to assign them in an orderly way to their causes. Hence, it calls upon nearly all other branches of science for help—astronomy, physics, chemistry, mineralogy, botany, and zoology. At the same time, it is essentially a recreative pursuit, for, as Mr. Horace Woodward says in his *History of the Geological Society of London*—published by the society —"the fulness of the science can never be attained without the vivifying influence of mountain and moor, of valley and sea coast." It is owing to this that the soldiers of the hammer, from Murchison, Sedgwick, Lyell, Ramsay, Etheridge, Salter, onwards to the present generation of "stone-crackers," are amongst the happiest, most genial, and mentally alert of our men of science.

That word "stone-cracker" I take from a letter addressed to me when I was a boy of twelve by the Rev. J. S. Henslow, Professor of Mineralogy and later of Botany at Cambridge, founder, with Adam Sedgwick, the great Woodwardian Professor of Geology, of the now flourishing Cambridge Philosophical Society, and the teacher, guide, and fateful friend of Charles Darwin. It was he who sent Darwin on the voyage of the *Beagle*. I had met this wonderful old naturalist at Felixstowe when exploring the marshes for rare plants and insects with my father. My father was a first-rate man at a country walk, and could tell you all the time about the flowers, flies, stones, and bones you might encounter. But Henslow surpassed him. I remember to this day nearly every word Henslow said, and everything he did on that memorable afternoon nearly fifty years ago. Amongst other things he explained how the rough flint implements recently discovered in river gravels—proving man's great antiquity—could be shown to owe their shape to blows, each blow causing a " conchoidal " fracture. And he struck with his hammer some very large flints which were lying in a heap in the meadow, and produced the most perfect dome-like broken surface or bulb of percussion. He promised to give me a real palæolithic flint implement and also a geological hammer. The letter which reached me later in London ran as follows: " Dear incipient Stonecracker—Enclosed you will find a draft for 10s. with which, at the shop in Newgate-street, you can obtain a geological hammer identical in all respects with my own. . . . In a separate parcel I send you a flint implement which I obtained myself in the gravel pit at St. Acheuil. . . ." The hammer, the flint-axe, and the letter are to this day treasured with deep affection and reverence for the giver, by the boy who was thus so kindly initiated in the " art and mystery " of Stone-crackers. Henslow died in 1861 at the age of 65. His daughter was the first wife of Sir Joseph Hooker, the great botanist and traveller, who celebrated his ninetieth birthday in July, 1907, and is still in full mental and bodily health and vigour.

8. *Experiments with Precious Stones*

A man of science cannot say a word about experiments with precious stones nowadays, but he is liable to be misunderstood and represented as having discovered how to make valuable gems out of dirt, or of enormous size, and in vast quantity. Last year the production of a few small crystals by the electrical decomposition of bisulphide of carbon was announced as something to affect the stock market instead of as a matter of interest to a few learned chemists. The crystals were supposed—erroneously as it turned out—to be diamond. We were also gravely told that a competent French chemist had discovered, and that the distinguished geologist, Professor Lapparent, had communicated the fact to the Academy of Sciences, that the radiation of radium acting on " corindon," or, as we should prefer to write it in England, " corundum "—a base, dull, colourless crystal—converts that dull substance into sapphires, rubies, emeralds, and topazes—and that the dealers attest the value of the precious stones so produced. This is really great nonsense, and arises from a little confusion in the use of the names of precious stones, and ignorance of what the substances indicated by those names are—defects which we cannot attribute to the French chemist, but must suppose to have " crept in " to the reports which crossed the Channel. Corundum is a colourless crystal, opaque or translucent. In chemical composition it is the oxide of aluminium—standing in the same relation to that light, white metal as rust or hematite ore does to the metal iron. It would not be at all astounding if by simple treatment we could convert corundum into sapphire or into ruby, since sapphire and ruby have precisely the same chemical constitution as corundum—are, in fact, only coloured varieties of corundum. Sapphire is blue, transparent

c 2

corundum; green and yellow "sapphires" are also common. The Oriental ruby is similarly only red, transparent corundum—like it only oxide of aluminium or alumina.

Diamonds are pure crystalline transparent carbon. Commonly they are colourless and transparent, but are sometimes black or white and opaque. Transparent diamonds are often found of a straw colour, rarely of a deep blue (the Hope Diamond), more rarely green (the Dresden Diamond), and rarest of all red.

If radium were really able (as some people have wrongly inferred from the French experiments) to change the chemical nature of corundum and convert it into topaz and emerald, the case would be very different from that of merely changing the colour of the corundum. What is to-day called "topaz" is a sherry-yellow crystal consisting of silicate of alumina and of fluoride of alumina. It turns pink when heated, and is also known of a blue colour and colourless. The topaz of the ancients from the coasts of the Red Sea is of a different chemical nature, and is now called peridote. Yellow corundum is sometimes wrongly called Oriental topaz, and the yellow-brown quartz crystals properly known as cairngorms are sometimes wrongly called Scotch topaz. So that the word "topaz" is used loosely as well as strictly, and confusion results. Emerald is widely distinct from corundum, sapphire, and ruby. It is a silicate of alumina and beryllium, and in its coarse and pale-coloured variety is known as beryl.

From all this it appears that some names of precious stones indicate substances quite distinct from one another chemically, built of differing elements, and also *per contra* that what is actually one and the same kind of precious stone in chemical composition and native crystalline form may present examples possessing various colours and degrees of transparency, each variety being called by a distinct name, and regarded popularly as a distinct kind of stone. Radium rays can convert colourless alumina or corundum into blue alumina (sapphire) or red alumina (ruby), but they cannot change alumina into beryllia (that is into emerald), nor into fluoride (that is into topaz).

One naturally asks, "To what is the colour of these precious stones due?" The answer is difficult, because very minute traces of chemical impurity, such as iron, cobalt, manganese, or chromium may suffice to tint an otherwise transparent, colourless crystal with the brightest red, yellow, blue, violet, or green. Moreover, it is certain from what we know of traces of metallic impurity in artificial glass that it may exist in such a state of chemical combination as to give no tint whatever to the glass, but after prolonged exposure to light or other agencies, the minute impurity may combine chemically with oxygen present in the glass and develop colour. Thus, for instance, old window-glass often assumes a violet or amethystine tint after long exposure. This varying colour of the combinations of metals according to whether they are oxidised or not, and the degree of oxidation, or the special salt which they may form, is in itself an unexpected thing to those who are not chemists. The metal chromium, for instance, gives rise to colourless, to yellow, red, green, and blue combinations. Manganese, a metal commonly associated with iron, gives rise to brilliant green, to violet, and to wine-red combinations, and if scattered as microscopic particles of black oxide in glass would produce no colour effect at all. From what we know of glass and the ease with which it is coloured to every shade of the rainbow by the admixture of traces of metallic impurities—so that "paste" or glass gems of all colours can be manufactured—it is not surprising to find that natural crystals, transparent and often devoid of colour (such as corundum, diamond, quartz, and topaz), are yet also found more or less frequently coloured in various tints. Nevertheless, it is the fact that in very few cases have chemists been able to prove by analysis what precisely is the cause of the colour in any given crystal or precious stone, although they may strongly suspect this or that as the colour-giving impurity. The actual quantity of a metallic impurity sufficient to give a tint is so excessively minute that the chemist finds it impossible to determine what it is by examining one small precious stone. He has not a sufficient bulk of material to operate on.

Having reached this point, we can see that such potent disturbing agents as the rays of radium—penetrating, a colourless, or faintly-coloured, crystal—may determine oxidation or other chemical combination within the crystal of traces of metal (iron, cobalt, manganese, chromium) already present there, and so give it an increased colour or an altogether new tint. In 1905 (therefore long before the recent French experiments had shown that the radium rays will act in this way on corundum, the " base variety" of sapphire and ruby), Sir William Crookes published an account of his experiments as to the action of the radium rays on the diamond. " Some fine colourless crystals of diamond," writes Sir William Crookes in 1905, " were embedded in radium bromide, and kept undisturbed for more than twelve months. At the end of that time they were examined. The radium had caused them to assume a beautiful bluish-green colour, and their value as 'fancy stones' had been materially increased." On another occasion Sir William found that a yellowish " off colour " diamond had its tint changed to a pale blue-green when embedded for six weeks in a tube with radium bromide. (I have seen this stone.) He also has succeeded in improving the clearness of diamonds by exposing them to radium rays. Everyone who has experimented with radium knows that it causes the glass which may be used to keep it covered to develop a brown or purple tint. This, then, is the explanation of the results obtained by the French observer with corundum, as reported a few months ago. There was no "transformation" of one substance into another, nor did he himself suggest that there was. The radium rays merely acted chemically on minute impurities present in colourless or pale-coloured crystals, and so produced colour as they do in diamonds or in glass.

9. Diamonds

His Majesty King Edward was presented with the great Cullinan diamond from the Transvaal in November 1907. This diamond weighs one pound and one-third (avoirdupois)—more than 21 oz. I have placed a good glass model of it in the Central Hall of the Natural History Museum; in the case with it is a glass model of another big diamond, the "Excelsior," as now cut, and also models of the "Pitt" diamond, in the rough and in the cut condition. Diamonds lose enormously in the process of cutting. The Excelsior, like the Cullinan, is a Cape diamond of fine quality, and free from colour. It was the biggest diamond known until the giant Cullinan was found: in the rough it weighed 7 oz., or less than a third of the Cullinan. As now cut, it only weighs $1\frac{3}{4}$ oz. It is reduced to a quarter of its original size.

In the same way, the Pitt diamond, an Indian one, named after General Pitt, of Madras, weighed originally 3 oz., and is now (it is in Paris, in the Louvre, and is called "The Regent") less than an ounce in weight. The biggest Indian diamond known—the Nizam—is not quite twice this size, whilst the Kohinoor, which is probably a fragment (a third) of the "Great Mogul"— a diamond which has disappeared, leaving only tradition and surmises as to its history—weighs no more than three-quarters of an ounce. This seems a small affair by the side of the twenty-one ounces of the Cullinan.

No one can guess what will happen to the Cullinan in cutting it. At the best, it may be reduced to something between four and five ounces in weight, and it may "fly" into fragments. It would be necessary deliberately to cut it up into smaller stones in order to obtain the full result of flashing of light and colour which twenty-one ounces of diamond can produce. And the operation of cutting and polishing is enormously expensive. One would have hoped that Sir William Crookes and other men of science would have been asked to examine this wonderful mass of transparent carbon by means of polarised light, Röntgen rays, and radium, and to determine exactly its specific gravity before it was broken up. Indeed, it would probably have retained its greatest interest and value if never cut at all.

Glass or "paste," as it is called, is made which cannot when new be distinguished from diamond by anyone but an expert, armed with the necessary tests. And the same is true as to paste imitations of all precious stones excepting the emerald (whose beautiful green tint cannot be exactly obtained), the cat's-eye, which has a peculiar fibrous structure, and the opal. The real value and quality of precious stones, as compared with glass, depends on their durability, their hardness, their resistance to scratching, and "dulling" of face and edge. Even our Anglo-Saxon ancestors, as may be seen in the fine collection recently dug up at Ipswich by Miss Layard, and placed in the old house serving as the municipal museum there, made gems of glass and paste. In modern times the art of making artificial "precious stones" has reached a degree of perfection which, so far as decorative purposes are concerned, leaves the natural stones no claim to superiority.

Gigantic as the Cullinan diamond is, it represents only about half the daily output of the De Beers mines. By the end of 1904 ten tons of diamonds, valued at £60,000,000 sterling, had been removed from the Kimberley mines. It is difficult to imagine what has become of them all, and since they are, unlike paste, durable and permanent, how the demand for additions to those in use, keeps up. Twelve years ago about four million pounds was spent annually by the public on the purchase of diamonds. It is stated that the annual demand and expenditure are now even larger.

Diamond is a peculiar form or variety of the chemical element carbon—a very peculiar form most people will say who remember that charcoal and lamp-black are the common form of carbon. That one and the same unchangeable chemical element can exist as an amorphous black lump or powder, and also without addition or loss of chemical constituents, as the clearest, hardest, and most brilliant of crystals, is a paradox. The same strange capacity for existing in two totally different forms is exhibited by other fairly familiar elements. Sulphur is found in tertiary water-deposited clays in Sicily (it has nothing to do with Etna or Vesuvius) in the form of clear, lemon-coloured crystals half an inch or more in length. If you take some commercial stick-sulphur and melt it in a porcelain spoon, and pour half the melted stuff like treacle into a jar of water, you will find that it cools as translucent threads which are pliable and soft. The other half which you leave in the spoon to cool shoots out into the form of long brittle crystals of a needle-like shape. These two varieties of sulphur are nearly as different as lamp-black and diamond.

Diamonds are found at the Cape in a " blue ground " which is of volcanic origin, formed by the action of steam under enormous pressure. The blue volcanic mud has been thrust up from great depths in the earth's surface in the form of " pipes " 100 yards to half a mile in diameter.   It has long been known that at very high temperatures (4,000 deg. Centigrade) the metal iron dissolves carbon.   The late Professor Moissan, of Paris, obtained artificial diamonds by suddenly cooling the iron in which carbon was dissolved by plunging the crucible into water.   The outer shell of iron cools and forms a tightly closed shell enclosing the still liquid core.   As this core cools it tends to expand, and thus produces an enormous pressure.   The melted carbon cooling under this pressure assumes the crystalline colourless form known as diamond.   There is good reason to believe that diamonds are formed, or have been formed, in association with metallic iron in a similar way, on a large scale, in great depths of the earth's crust, and are shot up to the surface with other débris in the volcanic steam mud which is the " blue ground."

A few diamonds of small size have been found in the Ural Mountains, otherwise they are not natural products of the northern hemisphere.   It is in India, Australia, South America, and South Africa that they are picked up, either in beds of streams, or in peculiar volcanic mud, or embedded in even harder rock.   Many are in a condition of severe strain when found, and contain minute cavities filled with liquid carbonic acid.   They are liable, in consequence, to break or even fly into powder when warmed by the hand or struck.   Though usually colourless, diamonds may be yellow, green, blue, or red, and the rays of radium cause colourless diamonds to become coloured.   Some diamonds, but not all, are phosphorescent—that is to say, like the well-known luminous paint—after exposure to strong light they acquire the power of shining themselves for a certain time when removed to a dark chamber.   And the curious thing is that, though themselves colourless, some give out blue, some green, some yellow, and some red light. The most wonderful, however, in this respect are the rare diamonds which become luminous merely by rubbing, and leave phosphorescent streaks on the cloth with which they are rubbed. This property is similar to the phosphorescence shown by other kinds of crystals when heated or when simply fractured.

Diamonds are readily distinguished from paste by the Röntgen rays, since they are transparent to those rays, whilst paste (or glass) is opaque to them. Radium also causes diamonds, but not paste, to phosphoresce. All diamonds are not equally hard, though they are the hardest of stones, and harder than steel, but not harder than the metal tantalum. Some Australian diamonds are known (from Inverel, New South Wales) which are so hard that at one time they could not be cut and polished; but only four years ago the rapidity of the wheels used in these processes was greatly increased, and these terribly hard diamonds were brought into subjection.

Thus it is clear that there are many extraordinary features of interest about the diamond, and that its brilliance and high price constitute only a small part of its fascination.

10. *Science and Fisheries*

Science, the knowledge of the vast system of orderly, inexorable activities under which we exist, and of which we, and all that we can apprehend, are but more or less significant parts, is not only to be regarded as a gratification of our curiosity, as food for our imagination, and the basis of our philosophical theories. It is, in addition to these, a thing of unparalleled importance to the immediate daily welfare of every man, woman, and child, and upon its due cultivation and use depend the future welfare, even the existence, of whole races of

mankind.   It is a startling fact that so few of those who
undertake to lead and to legislate for the people of this
country have any real conviction, or even a dim understanding of this truth.

In November 1906 a Committee appointed by the
Government took evidence as to the desirablity of
continuing the international investigation of the North
Sea, upon which Great Britain entered five years ago in
conjunction with other Northern States.   Only a few
weeks before, a number of scientific experts engaged in
this study of the North Sea, with a view to gaining
such knowledge of that great " waste of waters " as may
help the nations of adjacent lands to draw from it stores
of food without destroying the source or recklessly
injuring the supply, were entertained at dinner, at the
Guildhall, by the City Fathers, and treated to speeches
by hereditary legislators.   The view expressed by these
speakers was that the interests of the great fishing
industry and of the fish trade were best understood by
the practical fisherman.   Science was a " handmaid,"
useful in her place, but not to be permitted to under-
mine established interests and the hoary wisdom of the
practical man, her employer.   A German expert of
high official position, one of the guests, took a different
line.   He was astonished, even shocked, that Great
Britain, the State most largely concerned in the North
Sea fisheries, should be hesitating about continuing to
take part in the international investigation.   In Germany,
he said, they took a different course in such matters.
Men of business and practical legislators, when called
upon to deal with an important problem, sought first of
all for scientific knowledge of the conditions in question,
as complete and thorough as possible, and then
proceeded to act upon the sure foundation gained.
More knowledge, much more knowledge as to the causes
and conditions at work in regard to the life and movements of fishes in the North Sea was needed.   The work

of the International Committee must be continued, and his (the German) Government would certainly continue to do its share of the work.

The contrast in the British and the German attitude towards science is what is interesting in this episode. It is true that men of science in this country have to be content to take a very modest part in public affairs, and to allow politicians and self-styled " practical " men to treat science as " a handmaiden "—thankful when science is not regarded as an enemy. But they know well enough, and those who are really " practical men " know, that science is no handmaiden, but in reality the master—the master who must be obeyed; who alone can give true guidance; who alone can save the State. The sooner and the more thoroughly the people of this country have recognised this fact, and insist upon its unqualified acceptance in practice by their representatives and governors, the better for them and their posterity.

11. *Discoveries as to Malaria*

Recent scientific work, discovery, and application to practical affairs of the results of discovery, in regard to three great obstacles to human life and prosperity illustrate the vital importance to the state of scientific research. The obstacles in question are the diseases known as malaria, yellow fever, and Mediterranean, or Malta fever. It is now twenty-five years since Dr. Laveran, of Paris, discovered that malaria, or ague, is caused by a very minute parasite which exists in the red blood corpuscles of those stricken with the fever, and suggested that it is probably carried from victim to victim by blood-sucking mosquitoes (gnats). Major Ross, of the Indian Army, who has been rewarded for his discovery by the Nobel prize, determined to find out what gnat it is which carries the malaria-germ from man to man, and by most persevering experiment and microscopic examination showed that it is not the commoner gnat or mosquito (Culex), but the spot-winged kind (Anopheles), which alone can spread the malarial infection.   But Major Ross is, before everything else, a medical man, and his great purpose has been to apply his discovery to the prevention of disease.

Whole regions of the earth's surface are rendered dangerous, or even uninhabitable, for civilised men by malaria; in other words, by the Anopheles mosquito. Accordingly, Ross set to work to find the best means of destroying these agents of disease.   He found that the Anopheles gnat breeds in natural collections of water lying upon the surface of the ground in open country, and not as many common varieties of gnats do, in vessels and cisterns in houses.   The pools frequented by the malaria-carrying gnat are small and easily drained.   The obvious direction of science, therefore, was to remove or to cover up these pools wherever they were found in the neighbourhood of human habitations. Although Major Ross made his discoveries in India, and although he opened a campaign against malaria by removal of surface pools in the Colonies of West Africa—" the white man's grave "—twice visiting the chief British settlements—only half-hearted, incomplete measures have been taken, insufficient funds have been expended, and a supine executive and half-incredulous officials have failed to do more than partially reduce the prevalence of malaria in those regions.   On the other hand, where intelligent officials have understood and accepted the clear results of science in regard to malaria, the most striking and satisfactory consequences have followed.

At Ismailia, on the Suez Canal, malaria was almost universal; in 1866 there were in a population of eight thousand, 2,300 cases.   In 1897 there were over 2,000, and in 1902, when Ross was asked by the Prince d'Arenberg to visit the place and advise as to measures to be taken, there were 1,551 cases.   Ross directed the filling up of the breeding pools. The marshes were filled up with sand, the irrigation channels were deepened or treated with kerosene oil (which spreads as a fine film, and chokes the gnat larvæ), and the cess-pits were rendered uninhabitable by chemical treatment.   In one year the cases of malaria fell to 214, in 1905 they were only thirty-seven, and now the Suez Canal Company officially reports, "all trace of malaria has disappeared from Ismailia." The same satisfactory results have been obtained in Port Said, in Khartoum, in Port Swettenham of the Federated Malay States, in Havannah City, in Panama, and, in fact, wherever intelligent conviction has led to the active and complete employment of the methods necessary for the destruction of the gnats.   Under the British Government of India and the African and West India Colonies, little has been done.   Why? Because of the handmaiden theory and the ostrich-like refusal of our officials to face and accept the master.

An even more wonderful and beneficent result has been obtained in the case of that terrible disease " Yellow Jack," or " Black Vomit "—the yellow fever. Owing to the discoveries and definite proof by Ross as to the part played by gnats in malaria, the able medical men in the public service of the United States of America have thoroughly examined experimentally the mode of infection of human beings with the germ of yellow fever, and have conclusively proved that infection is solely and entirely due to the bite of one species of gnat—the Stegomyia fasciata.   They have proved to absolute certainty that yellow fever is not carried through the air, nor by food or drink, nor by contact with infected persons or their cloths or emanations, but only by the fasciate gnat, a house-frequenting species, which sucks the blood of a yellow fever patient, and after twelve days, and not till then, becomes capable of imparting the infection to those whom it may stab or "bite." The firm demonstration of this fact was not made without great devotion, courage, and self-sacrifice. In the ardour of their pursuit not a few of the experimenters risked and lost their lives. Among these the name of Dr. Lazear, of the United States Army, is prominent. He deliberately permitted himself to be bitten by a stray mosquito in a yellow fever hospital, in order to show that the insect could convey the infection. He was bitten on Sept. 13, 1900, and died on Sept. 25, having proved his point.

The actual germ, microbe, or minute parasitic organism which causes yellow fever, and is carried by the fasciate gnat, has not yet been detected. Nevertheless, without seeing and isolating the microbe, the medical men of America (Sternberg, Finlay, Carroll, and others) have, by destroying the gnat and preventing its access to men—especially to patients already infected, and, therefore, certain to infect the gnats and cause them to spread the disease—practically made an end of yellow fever in many great cities of the New World, where it was only six years ago an ever-present horror, striking men down with a suddenness and with a deadliness which paralysed human activity. Here, as in other cases, intelligent appreciation of the results of science by a governor or a municipality has saved thousands of lives. On the other hand, in Rio de Janeiro, "the opposition encountered by the sanitary authorities of the city from political factions and the ridicule to which they were subjected by the local Press" were insuperable (I quote from an official report), and so a few more thousand lives were sacrificed before the master was recognised and the proffered safety accepted. In Vera Cruz, in New Orleans, and in Panama yellow fever has been reduced to a vanishing quantity by removing the pools and tanks in which the fasciate gnat can breed, and by making use of wire-gauze to prevent the access of mosquitoes to houses, bed-chambers, drains, and baths, and especially to prevent not only their access to, but their egress from, the rooms and beds of patients already infected with disease.

In the city of Havannah, during the American occupation of Cuba (1900-1903), Colonel Gorgas reduced the death-rate due to yellow fever from an annual average of 751 to so small a figure as six. The same energetic and faithful administrator has been at work, with even more remarkable results, in the canal zone of the Isthmus of Panama since 1904. The attempt of the French to cut the canal was foiled chiefly by yellow fever and malaria. It is estimated that their effort cost quite 50,000 lives. Assisted by an able and enthusiastic staff, and charged with the task by a Government which comprehends the fact that the really " practical men " are the men who recognise science as the master (not as the negligible eccentric handmaid), Colonel Gorgas has banished the mosquito from his zone of occupation. As a consequence there is neither malaria nor yellow fever on the Panama works. In 1906 the total death-rate amongst 5,000 white employés on the Panama Canal works was only seven in the thousand. Further, in last April the daily sick-rate of the total force of about 40,000 people was only seventeen in the thousand. Colonel Gorgas declares that there is but little sickness of any kind among the Americans in the employ of the Panama Commission, and that they and their wives and children are fully as vigorous and robust in appearance and in fact, as the same number of people in the United States. There is no reason why the centres of wealth, civilisation, and population should not again be in the tropics, as they were in the dawn of man's history.

12. *Malta Fever*

Mediterranean or Malta fever was for long confused with typhoid and others fevers. Our soldiers and sailors at Malta, Gibraltar, and Cyprus, as well as many frequenters of the African and Asiatic shore, were subject to this disease, and often incapacitated by it. In 1887 Colonel David Bruce discovered in the blood of patients the minute Micrococcus melitensis, which is its cause, and established the fact that it is a definite independent disease. The hospital at Malta has received as many as 624 patients in a year suffering from Malta fever from among the 8,000 soldiers on the island and the 12,000 sailors on the Mediterranean Station. And as they stay in hospital on an average for four months, this means 74,880 days of illness. This means a considerable loss to the State, as well as a large amount of personal suffering terminated, in some cases after two years' sickness, by death.

The War Office, Admiralty, and Colonial Office applied in 1904 to the Royal Society of London to undertake a further investigation of this disease. The society sent out a small commission, which has been at work for three years, and has published seven volumes of reports. The problem before the commission was to discover the mode of infection by the Malta-fever germ (the Micrococcus melitensis), and thus, if possible, to arrive at a means of arresting the infection. Various hypotheses, guesses as to probable and possible methods of dissemination, were entertained and examined. As the germ occurs in the blood, it was naturally considered possible that gnats or other insects were the carrying agent. But negative results followed all experiments in this direction. Then it was found that the " germ " passes out of the body in large quantities by the renal secretion, and it was thought that it might be conveyed in a dried form with dust in the air. This also proved to be an incorrect supposition.

Next a very important discovery was made. The germ was found in the blood and the excretions of 10 per cent. of the goats which are kept in Malta as the sole source of milk, and are driven through the streets to supply customers, whilst 50 per cent. of the goats were found to have been infected at some time. Then the germ was found in the milk itself, and it only remained to prove by experiment that it was from the goats' milk that human beings acquire the infection. A monkey fed with the milk of an infected goat acquired the fever.

The next step was to stop the consumption of goats' milk by the soldiers and sailors in the hospital and barrack. Actually we were carefully feeding our invalid soldiers and sailors in the great hospital at Valetta with a highly poisonous infected fluid—the milk of the Maltese goat! The preventive measure—the stoppage of goats' milk—only came into operation in July, 1906. In the first six months of that year there were thirty-one cases of Malta fever in every thousand of the garrison (numbering about 8,000 men). In the preceding six months there had been forty-seven cases per thousand. Now when the goats' milk was stopped after July, 1906, what was the result? From July to December, 1906, there were only ten cases per thousand of the garrison. In actual numbers there were in July, August, and September in 1905 as many as 258 cases, whilst in the same months in 1906, after removal of goats' milk from the dietary of the troops, there were only twenty-six cases, and these were probably due to the independent purchase of goats' milk by soldiers outside the barracks. In the naval hospital until 1906 almost every patient who remained in the hospital a few weeks took the disease. Since the exclusion of goats' milk not a single case has occurred.

The Director-General of the Medical Department of the Navy reports that there has been no case of Malta fever during the year among the sailors, and only seven cases among the soldiers up to the end of September, 1907.

Gibraltar had a fever of its own, identical with Malta fever. It has now been shown that it was probably introduced by the importation of goats from Malta for the supply of milk. This is likely, because the importation of Maltese goats ceased in 1883, and the fever began to disappear from Gibraltar in 1885, and finally vanished altogether in 1905.

In South Africa Malta-fever is common amongst the white population. It is probable, according to Colonel Birt, that it was introduced by means of infected goats imported from the Mediterranean. The soldiers, however, in South Africa are free from this disease, excepting those who have already contracted it in the Mediterranean, since in South Africa goats' milk does not enter into the dietary of the soldier. It is the civilian population which suffers.

13. *A Cure for Sleeping Sickness*

Diamonds and sleeping sickness are both special African problems. It was owing to the proposal to employ natives from Uganda in the South African diamond mines that the Colonial Secretary (Mr. Chamberlain at that date) asked the Royal Society to say whether the sleeping sickness which had broken out with terrible violence in Central Africa constituted an obstacle to that employment, on account of the danger of introducing the disease into South Africa. The Royal Society advised the Government not to allow the transport of natives from the infected districts of Uganda, and sent out a commission to Central Africa to study the disease. The result was the discovery by Colonel Bruce of the parasite of sleeping sickness called Trypanosoma—a kind previously known in some other diseases—and of the fact that it is a tsetse-fly which carries it. A quarter of a million natives have died in Central Africa within the last six years from sleeping sickness. The Tropical Diseases Committee of the Royal Society has started an inquiry into the action of drugs on the parasites (known as trypanosomes) which cause sleeping sickness and the horse and cattle disease of the " fly-belts " of South Africa.

The minute parasites which cause Malta, yellow, and malarial fever, and other infections, are no doubt best dealt with by excluding them from access to the human body when that is possible. But once they have effected a lodgment and commenced to multiply in the blood or tissues, it is still possible to get at them by means of drugs, which poison them without injuring their human victim. Thus quinine has been of enormous service in checking the ravages of the malaria parasite, and really in Great Britain has exterminated " ague," which is the English name for malaria. Many experiments have been made during the last two years, with the view of finding some drug which will, in like manner, destroy the trypanosomes which have established themselves in the blood and lymph-passages of the human body, and are slowly killing their victim with sleeping sickness. An arsenic compound, " atoxyl," has been found effective when injected into the patient's body, and according to Dr. Koch, who returned last year from Uganda, he has found nothing better than this treatment, discovered by Dr. Thomas and Dr. Breinl, of the Liverpool School of Tropical Medicine, three years ago. Dr. Plimmer and Dr. Thomson, who have been experimenting in London for the Royal Society, have found a drug which is more effective than atoxyl in destroying certain trypanosomes which attack rats, and is now being tried in the treatment of sleeping sickness. This is the tartrate of sodium and antimony—a salt corresponding to the well-known tartar emetic, with this difference, that it contains sodium instead of potassium. It seems that this sodium variety of tartar emetic is very destructive to trypanosomes in the blood and lymph, and has no injurious effect of a lowering nature, such as occurs when the potassium salt is used. As the antimony drug is far cheaper than atoxyl, it will be possible to apply it freely to horses and cattle suffering from " nagana " and " surra," which are diseases due to trypanosomes of a special kind. Two white men who had become infected by the trypanosome of sleeping sickness in West Africa have been treated with the new drug in London, 'and the parasites have completely disappeared from their blood in consequence, though it remains to be seen whether a permanent cure has been effected. One cannot imagine a situation of more thrilling interest than that existing in the nursing home where those two victims were given a strong hope of escape from what seemed to be certain death, whilst the fate of thousands of African natives, similarly infected, was hanging in the balance ! After six months from the date of treatment the report is satisfactory. The parasites have not yet re-appeared (July, 1908) in the two patients treated in November.

14. *Tsetse-Flies and Disease*

Dr. Koch appears to have been questioned on his return to Europe by some journalists as to the results of his study of sleeping sickness during the past year and a half in Uganda. It was already known (three years ago), from the observations of Professor Minchin, Dr. Gray, and Dr. Tulloch (the Royal Society's observers in Uganda), that the tsetse-fly in Uganda sucks the blood of crocodiles, also of fishes and of hippopotami. Dr. Koch confirms this observation. Minchin also observed a trypanosome in the blood of the crocodile differing from that of sleeping sickness. Whether crocodiles help, in an important degree, to keep tsetse-flies alive in the regions where they occur, by offering them a ready meal of blood, is uncertain. So far as the facts are known, they do not lead to the belief that the crocodile is a " reservoir host " for the trypanosome of sleeping sickness.

" Reservoir-host " is a very useful and expressive name for animals which can tolerate or support a parasite in their blood which is deadly to other animals. The parasite flourishes in abundance in the reservoir-host with entire satisfaction to both host and guest. But a blood-sucking fly or gnat, of promiscuous tastes in the matter of blood, comes along, sucks the reservoir-host a bit, and then goes off for another meal to a susceptible animal, into which it introduces the parasite now adhering to its already blood-smeared proboscis or beak. Such a history was first established by Bruce in regard to the trypanosome parasite which causes the deadly nagana disease in the " fly-belts " of South Africa. The big game animals are reservoir-hosts to this parasite, from which they are carried by the tsetse-fly to horses, mules, and dogs, which, being of foreign origin, are not tolerant of it, but are killed by the poison to which its multiplication in their blood gives rise. Thus, too, native children, both in Africa and the East Indies, appear to be tolerant of the malaria parasite, and act as reservoir-hosts from which the spot-winged gnats suck and distribute the parasite to the non-tolerant, susceptible adult natives and white men.

The tsetse-flies are little bigger than thc common house-fly, and bite, or rather stab, very rapidly after alighting on the skin. The study of flies and gnats, and

other blood-sucking insects, has become extremely
important, and has been carried on with great energy by
many specialists since it became known that these insects
play such a terribly important part in the causation
of disease.   At the Natural History Museum I received
(in response to a circular issued at my request by H.M.
Government) thousands of specimens of gnats (mosquitoes) from all parts of the world, and some hundreds of
new species have been described in a series of volumes by
Professor F. V. Theobald, published by the trustees.
Other volumes are in preparation illustrating the blood-
sucking flies of various regions of the world, and one
concerning those of the British Islands has already
appeared.   The common gnat, the spot-winged gnat,
and the tsetse-fly—as well as the microscopic parasites
causing malaria and sleeping sickness—are illustrated
by greatly enlarged models—very carefully executed
under my direction, which are exhibited in the central
hall of the museum.

It is a curious fact that the coloured races of men—
especially those of Africa—have little or no objection to
being bitten by flies.   They seem to accept the attention of flies and ticks with indifference.   The men sleep
in the day under trees, and are willing food-supply to
the insects.   The eyelids of children are literally
inhabited by flies in some countries, and the folds of
the skin of fat adults hide whole rows of fast-holding
ticks.   But the white man does not willingly permit
either fly, flea, or gnat to settle on him.   He is (or has
been), nevertheless, unwisely tolerant of house-flies in
his habitations, and the poorer and less cleanly population are in large proportion infested with wingless
insects.   The newly established knowledge that certain
flies (glossina or tsetse-fly) are the carriers of sleeping
sickness, that gnats are the carriers of malaria and of
yellow fever, that fleas are the carriers of the plague,
and that certain kinds of ticks are the carriers of cattle-

fevers and dog-fevers, and probably of some obscure
fevers of man, must make us all more anxious than we
were about contact with insect life. For ages popular
tradition has ascribed diseases of one kind and another
in various parts of the world to the bites of flies. But
actually it is little more than fifty years ago since it was
really shown that deadly germs or parasites existed
which could be, and actually are, carried by flies from
one animal to another, and introduced into the blood by
the flies' stab. This was first shown in regard to the
bacterium of splenic fever (or anthrax, or wool-sorters'
disease), a blood-disease of cattle which is transferred
by the big, fiercely-biting " horse-flies " (tabanus), from
animals to man, and is invariably fatal. Another
bacterial disease, " pernicious œdema," is inflicted on
man in the same way. These cases were exceptional, and
it is only quite recently that the agency of flies and fleas
in great epidemics, and in diseases causes thousands
of deaths every year in well-known regions, has been
discovered.

15. *Monkeys and Fleas*

The wingless parasites known as pediculi are not
known as active agents in spreading disease germs, probably because they do not readily transfer themselves
from one animal to another. It is in this connection a
really remarkable fact that monkeys are not infested by
fleas, and that only in few cases and not in many kinds
have pediculi or acari been observed. In this respect the
lower races of men (and even the higher) seem to have
fallen away from a grade of excellence attained by their
despised quadrumanous cousins. When this fact as to
the freedom of monkeys from insect parasites is mentioned, those who have watched monkeys in captivity
will immediately say, " Surely I have seen monkeys

carefully picking insects from one another's fur." The
fact is that it is this very habit of "picking" which
prevents monkeys from harbouring fleas. Whereas a
dog or a cat can only scratch, the monkey has an
opposible thumb and delicately sensitive fingers. That
which has become the hand of man, with all its marvellous skill and efficiency, has been elaborated in its early
stages as a means for keeping the hair clean. When
monkeys are seen carefully removing something with finger
and thumb from their own or their companion's hair, it
is not an insect but a little piece of fatty secretion and
scurf which is thus removed. The habit, which seems to
be general in all kinds of monkeys, even with the
anthropoids, such as the chimpanzee and the orang, has
of course been efficient in removing any parasitic insects
which may at one time have infested monkeys—all
other furry animals are liberally supplied with them, as
also are birds—but is now preventive of any re-establish-
ment of such visitors. The popular judgment of the
monkey's habit is similar to that of the Japanese Aino,
who remarked to a traveller who arranged to have a
bath in his room every day that he must be a very
dirty man to require it.

16. *The Jigger Flea*

One flea is recorded as having been once taken on
an anthropoid ape (a gorilla), and is the "jigger," Pulex
penetrans. This is a very serious pest, the history of
which shows how man himself opens up the path by
which dangerous diseases spread. The jigger-flea was
originally known only in the South American tropics.
It spread from there to the West Indies in the last
century. It burrows into the skin, usually between the
toes, but elsewhere also, and causes an abscess and sore
as big and deep as a hazel-nut. Several such cavities at

a time are dangerous, and often lead to blood-poisoning and death. Europeans avoid the burrowing of the jigger by having their toes carefully examined every morning, but black men are less careful. From the West Indies, about thirty years ago, the jigger was carried in ships to West Africa. There it flourished and spread from village to village across Central Africa, decimating the population. It appears to have been carried to a large extent by dogs, in whose skin it flourishes. It has now passed through Africa to India, and we shall no doubt soon hear of its having completed the circuit of the globe.

A great many kinds of fleas are known, many furry animals having their own special species, which does not leave them to take up its dwelling on other kinds of animal. The common rat has a large flea of its own, which apparently is not the flea which carries the plague from rats to men. It is a " wandering " flea which does this, namely, the Cheops flea. This flea, common in the East but unknown in colder regions, does not stay as one could wish it to do—on the rat ; but travels about visiting human beings and dogs, and so carries the plague bacillus from rats to men. In the absence of these fleas plague would be a rat-disease unknown in men. It is probable that we do not nowadays live so thoroughly cheek-by-jowl with rats in Western Europe as formerly, so that even if rats infected with plague and harbouring the Eastern Cheops flea arrive in our docks, the wandering flea is too far off to reach us in our modern houses.

17. *Public Estimate of the Value of Science*

The Royal Society, the full title of which is The Royal Society of London for the Promotion of Natural Knowledge, has its anniversary meeting and dinner on

St. Andrew's Day. The health of the medallists of the year 1907 was given from the chair by Lord Rayleigh, and they replied one by one to the toast. Professor Michelsen, of Chicago, received what is considered the greatest honour the society has to bestow—the Copley Medal (founded more than two hundred years ago) for his researches on light. He related in his speech how he had tried to interest a wealthy business man in the experiments going on in his laboratory, in the hope that his friend might be moved to give pecuniary aid for the provision of new apparatus. One by one, he showed his delicate instruments and explained their uses; no impression was produced. At last he explained how the bright lines of the spectrum of flame, coloured by incandescent elements (such as theatre-goers know as red fire, green fire, blue fire, &c.), can be recognised by means of the spectroscope in the light of the sun— proving the presence of the metals and other elements of this earth in that remote body. He especially explained and showed his friend the experiments by which sodium, the metal of which caustic soda is the "rust," is thus proved to be present in the sun. At last his friend spoke. He said: "Who the —— cares if there is sodium in the sun?" Professor Michelsen did not tell the fellows of the Royal Society how he replied to that abrupt inquiry.

A more encouraging speech was that of Lord Fitzmaurice, the Under-Secretary of State for Foreign Affairs, who replied to the toast of the guests. He declared, in so many words, " It is every day becoming more and more certain that science is the master." He said that in his own business as a diplomatist he found that the chief matters which he had to discuss and decide depended on scientific knowledge and the information and guidance given to him and his colleagues by scientific men. In the beginning of the eighteenth century the British Government had sent a bishop and a poet to negotiate the Treaty of Utrecht. But neither would be of any use in modern diplomacy. What they always had to seek at the present day was the aid of the scientific departments of the Navy or the Army, or of the Royal Society. Such matters as the relative merits of a Channel tunnel or a Channel ferry, the limitations of territory by land, by sea, or above the land in the air, the international agreements as to measures for checking the spread of disease or of insect pests, and, indeed, most matters which had come before him since he had been in office, had to be decided by the scientific experts. He did not propose that diplomatists should at once vacate their posts and endeavour to secure the occupation of them by men of science, but he thought that at no distant date such a course would be considered not only reasonable, but necessary !

18. *The Common House-fly and Others*

The common house-fly is not so innocent as he looks, but really a dirty little thing. He has not a sharp beak-like proboscis, and cannot stab, but he has a soft, dabbing proboscis, which he pushes on to every kind of filth as well as walking with his six legs on such matter. Then he comes and wipes off minute particles and germs on to our food, our lips, our fingers, and faces. It is quite certain that he, and others allied to him, are thus the means of spreading typhoid fever in camps where there are open latrines and open larders and mess tables. The house-fly breeds from a maggot, just as the blue-bottle or blow-fly does, but very few people have ever seen or recognised the maggot of the house-fly. The reason is that it lays its eggs in horse dung, and the grubs are hatched in the muck-heaps of stables. That is also the reason why it is much less numerous in

London than it used to be, since stables and mews are now fewer and cleaner than they were. It is also the reason why the house-fly abounds in ill-kept country inns and farmhouses. Its breeding ground is just outside the window.

There is not only one common house-fly in this country: there are three kinds, in addition to the blue-bottle or blow-fly, which is distinguished at once by its great size and blue colour, and lays its eggs in carrion. Late in the year you may often see what would pass for young or starveling house-flies going about among the others. This is a distinct species, the Homalomyia canicularis of entomologists. The third kind only to be distinguished by careful examination with the aid of a magnifying glass, is Anthomyia radicum. Both these are much less abundant than the common house-fly (Musca domestica), with which they almost always occur. Their breeding habits are similar to those of the common house-fly.

A fourth kind of fly is invariably mistaken for the common house-fly when it is noticed, as it sometimes is, in consequence of the sharp stab which it inflicts. As recently as the beginning of November last year I was "bitten" or pricked by one of this fourth kind in a London club. They are common enough on the sea shore in autumn, and may be a severe nuisance. People generally take them for common house-flies which have lost their temper in the hot weather and give way to the bad habit of "biting" out of sheer exasperation. Really, of course, a house-fly could not stab or prick with its broad-ended proboscis. The fly in question, which looks almost exactly like a well-grown house-fly, but possesses a sharp and business-like beak or proboscis, is known to scientific men as Stomoxys calcitrans. There are man kinds of Stomoxys scattered all over the world, andy it is probable, though not actually proved, that they carry parasites such as the trypanosomes of horse and cattle diseases from one animal to another, as do the species of Glossina or tsetse-fly.

But we have yet to learn more about these flies and the parasites they transfer. In the case of the gnat, it has been discovered that the malaria parasite is swallowed by the gnat, and multiplies in it, producing thousands of spores in its blood, and it is these spores which the gnat hands or rather "mouths" on to man. No such multiplication of the trypanosome in the tsetse-fly (Glossina) is known. The tsetse-fly passes on the trypanosome as it received it, and yet it seems as though it is not any and every biting fly which can pass on the trypanosome of nagana, or of sleeping sickness, but only the particular species of tsetse-fly. Perhaps it is a case of greater abundance, the tsetse-flies being the obvious and dangerous carriers of trypanosome disease where they occur, on account of their abundance and the fierceness and celerity of their attack. It is almost certain that in India, Burma, and South America some other flies must transfer the trypanosomes from animal to animal, causing the diseases known as surra and mal de caderas, because no tsetse-flies—that is to say, no flies of the genus Glossina—occur in those countries, and no other mode of transference, except by some blood-sucking insect, seems probable.

Ants in Africa are carriers of infection, and possibly also in London kitchens, where a little red ant sometimes abounds. The black beetle or cockroach is a creature to be got rid of, as it is very probable that it spreads certain kinds of infection over food and dishes during the hours of "revelry by night" which kind-hearted people allow it to enjoy in their kitchens.

19. *Cerebral Inhibition*

The best golf-player does not think, as he plays his stroke, of the hundred-and-one muscular contractions which, accurately co-ordinated, result in his making a fine drive or a perfect approach ; nor does the pianist examine the order of movement of his fingers. His "sub-liminal self," his "unconscious cerebration," attends to these details without his conscious intervention, and all the better for the absence of what the nerve-physiologists call "cerebral inhibition"—that is to say, the delay or arrest due to the sending round of the message or order to the muscles by way of the higher brain-centres, instead of letting it go directly from a lower centre without the intervention of the seats of attention and consciousness. The sneezing caused in most people by a pinch of ordinary snuff can be rendered impossible by "cerebral inhibition," set up by a wager with the snuff-taking victim that he will fail to sneeze in three minutes, however much snuff he may take. His attention to the mechanism of the anticipated sneeze, and his desire for it, inhibit the whole apparatus. So long as you can make him anxious to sneeze and fix his attention on the effort to do so, by a judicious exhortation at intervals, he will not succeed in sneezing. When the three minutes are up, and you both have ceased to be interested in the matter, he will probably sneeze unexpectedly and sharply. I was set on to this train of thought by a recent visit to an exhibition of photographs.

There were many very interesting illustrations of the application of photography to scientific investigation. Among others I saw a fine enlarged photograph of the common millipede (Julus terrestris), and my desire was renewed to have a bioscopic film-series of the movements of this creature's legs. Some years ago I attempted to analyse, and published an account of, the regular rhythmic movement of the legs of millipedes. I found that the "phases" of forward and backward swing are presented in groups of twelve pairs of legs, each pair of legs being in the same phase of movement as the twelfth pair beyond it. But instantaneous photography would give complete certainty about the movement in this case, and in the case of the even more beautiful "rippling" movement of the legs of some of the marine worms. Some kindly photographer might take up the investigation and prepare a series of films. The problem is raised and the effects of "cerebral inhibition" described in a little poem which I am told we owe to the author of "Lorna Doone." As it is not widely known, I give it here as a record of "cerebral inhibition":

> " A centipede was happy 'til
> One day a toad in fun
> Said, 'Pray which leg moves after which?'
> This raised her doubts to such a pitch
> She fell exhausted in the ditch,
> Not knowing how to run."

The point, of course, is that she could execute the complex movement of her legs well enough until her brain was set to work and her conscious attention given to the matter. Then "cerebral inhibition" took place and she broke down.

20. *Colour-photography and Photographs of Mars*

There were admirable photographs of wild birds and their nests, and of insects and plants in this exhibition. I saw the new Lumière coloured transparent photographs thrown by a lantern on the screen, and could distinguish the dots of red, green. and violet colour on what, at a little distance, appeared to be a brilliantly white part of the picture (the shirt collar of a "sitter"), just as one

E

sees a mosaic of coloured dots in the blazing sunlight of
the pictures painted by the French school of so-called
" vibristes " (Monod and others).   Perhaps the most remarkable of these photographs was a set of prints from
untouched photographs of the planet Mars, executed in
July 1907 by Professor Perceval Lowell at his observa-
tory in Arizona.

The Mars photographs are each about as big as a
dried pea (that is the biggest size possible with the
feeble light reflected by Mars), but "several of the
canals," says Mr. Lowell, " are distinctly visible on the
photographs, and one has been photographed double."
I should have liked to examine these photographs in a
good light with a lens.   The statement quoted means
that the canals in Mars can no longer be regarded as due
to errors of eyesight and imagination, and that the
annual doubling or formation of a second canal parallel
to what was earlier in the year a single canal, is actually
recorded by a disinterested, impartial photographic plate.
Are these canals the work of intelligent inhabitants of
Mars ?   I will not venture to say in reply more than
this, that I have never heard any other explanation of
their occurrence.   But that, of course, still leaves the
matter open.

21. *Origin of Names by Errors in Copying*

A curious illustration of a mistake perpetuated by a
clerical error is the title of Viscount Glerawly.   The
title was intended to have been Glenawly, but the bad
writing of a clerk converted the " n " into an " r," and
the name having been so entered in the patent of
nobility, or some such document, could not be altered.
The same thing has happened to the mammoth.   His
proper native name is " mammont," but " mont " became

"mont," and then "moth." A similar clerical error is responsible for the name Gavial, which is applied to the long, narrow-nosed crocodile of India, both as a scientific name (Gavialis) and colloquially. Really the "v" is due to a misreading of an "r," the creature's native name being Garial. It was so written down and sent home by an early explorer, but his handwriting being wanting in clearness, the word was copied as Gavial and the scientific patent issued in that name.

22. *False News as to Extinct Monsters*

The tendency of English newspapers to bedeck themselves every now and again with rank absurdities copied from American rubbish-sheets is a disease. On no subject outside the field of natural history and medicine would any editor dream of printing the stuff which does duty as "news" in regard to these departments— stuff which has not even the semblance of being carefully concocted, but yet is found "good enough" to circulate as serious information.

Another antediluvian monster, much larger than the mammoth, was reported in a London evening paper at the end of November 1907. The article devoted to it is a mass of absurdity, a burlesque of a genuine note on the subject. It appears that the most ordinary thing happened at Los Angelos, California, namely, that some workmen, in driving a tunnel, unearthed some fossil bones. We are not surprised to learn (though it is announced as a marvel) that the bones were those of a mastodon (of which you may see a whole skeleton in Cromwell-road), and those of the extinct American elephant called Elephas columbi. This very commonplace occurrence was certainly not worth recording in a London daily paper. So it is elaborately dressed up with details intended to "fetch" the innocent reader. The writer says Elephas columbi is as much larger than the Siberian mammoth as that is larger than the horse of to-day. The truth is that Elephas columbi and the mammoth are as nearly as possible of the same size. To writer goes on to tell of a "fossil horse," found at the same place, "a wonderful two-toed animal marked by his cloven hoof." That is cool impudence ; it is precisely "the double hoof" which none of the horse tribe possess, but all the deer, cattle, and sheep do. He next tells us that elephants and mastodons were never found together before, but supposed to have shunned each other's company. This is an invention ; their remains are found side by side all over Europe. Then suddenly the surprising statement is made, like a bolt from the blue, "England ceases to be the Mother Country and Germany the Fatherland to us," and the pre-eminence of America in providing the biggest thing on earth is declared to have been already manifest "when the world rose out of chaos." It is satisfactory to be told that England is not the Mother Country of this silliness ; but whether the world which solemnly prints and reads it can be said to have yet "risen out of chaos" must be regarded as doubtful.

23. Mistletoe and Holly

Christmas things and customs comprise much that has great interest from a scientific point of view. Our modern celebration of Christmas in England is a combination of the Christian festival of the Nativity with that of the Epiphany, and that of St. Nicholas, who long ago was substituted for the sea god Neptune, of classical mythology, by sea-faring folk. Santa Claus —or Saint Nicholas—has his festival at the beginning of December, but he has been carried over to Christmas Day, and appears as "Father Christmas" in modern celebrations. There is no great antiquity about this part of the tradition which we try to keep alive at Christmas. The making of Christmas Day and Christmastide into a special children's festival is, on the other hand, a moving back of the festival of the Epiphany, when gifts were brought to the child Christ by wise men of the East. In Rome I have assisted in celebrating our Twelfth Night under the name "Befani," at a great illuminated public fair, near the Pantheon, where children are taken to buy toys.

There has been in England also a similar moving back of the very ancient—even prehistoric—celebrations of the New Year to Christmas, and hence it is that the mysterious and sacred "mistletoe" of the Druids is mingled in our houses with the less significant but beautiful holly as a decoration. The Christian Church, however, did not, and does not, sanction the introduction of mistletoe into the sacred edifice, and not many years ago those who loved and truly understood tradition would not permit mistletoe to be mixed with holly even in the private house at Christmastide. Mistletoe, it was held, could not be rightly introduced until the new year. The new year, however, of the Druids differed in date from that of the later calendar, and fell in what is to us the second week of March.

The holly tree, with its splendid red berries and shining, prickly leaves, is a beautiful decorative plant, very hardy and abundant: it was used by the old Romans in their "Saturnalia," a feast which nearly coincided with the Christmas of the new religion. There is a species of holly in South America the leaves of which are made into tea by the Indians, the Paraguay tea or matté. This tea is an unpleasant, bitter decoction, devoid of aroma, if I may judge from samples which I have tasted in London. "Ilex" is the botanical name of the genus to which both our holly-tree and the Paraguay tea belong, but it must not be confused with the evergreen oak to which the name Quercus ilex is given on account of the resemblance of its leaves to those of a holly.

The mistletoe (or mistil-tan, the pale branch, in Anglo-Saxon) is a pale-coloured, small-flowered member of a great family of parasitic plants, the Loranthaceæ. They all live upon trees, and draw a part of their nourishment from the juices of the tree into which their rootlets penetrate. The tropical allies of the mistletoe are very beautiful plants, with fine bunches of brilliantly-coloured flowers and broad handsome green leaves. Our mistletoe is most commonly found parasitic on apple trees and poplar trees. It occurs on nearly all our trees, but is very rare on the oak. A careful · inquiry some time ago resulted in the discovery of only seven oaks in all England on which mistletoe was growing. The Druids took their sacred mistletoe from the sacred oak tree on account of its rarity. To them it was a charm against infertility and sterility, and, according to Pliny, was cut and distributed at the new year with great ceremony and the sacrifice of heifers. Its paired white berries contains a viscid fluid which gives it its botanical name Viscum album—and causes the seeds to adhere to the beaks of birds—and thus to be transported to a distance and introduced by the birds' attempts to wipe their beaks into the cracks of the bark of trees, in which the seeds germinate.

The white-berried mistletoe is the only English kind, and red mistletoe seems altogether out of character. But a red-berried species (Viscum cruciatum) is parasitic on the olive tree in Spain, North Africa, and Syria. Curiously enough, though the white-berried mistletoe is excommunicated by the Western Christian Church on account of its use in pagan worship, the red-berried mistletoe was gathered from olive trees in the Garden of Gethsemane and in the enclosure of the Holy Sepulchre at Jerusalem by Sir Joseph Hooker, the great botanist. The red-berried mistletoe was successfully raised from seed on young olive trees six years ago in this country by the Hon. Charles Ellis, of Frensham, near Haslemere, and was figured at that time by Hooker.

The mistletoe has an evil name in Scandinavian mythology. Baldur, the beautiful, the Sun-god, was made, like Achilles, invulnerable to spears and arrows cut from whatever tree grows on earth. All things had taken an oath not to hurt him, and the gods of Walhalla amused themselves by throwing all sorts of darts and clubs at him—none could hurt him. At last the blind god Höder, who loved the beautiful Baldur none the less because he himself was weakly and sightless, also ventured to throw a dart at his invulnerable friend. It sped home, pierced Baldur's heart, and killed him. The dart was made of mistletoe, a tree that does not grow on earth, but lives as a parasite high up on other trees, and had taken no oath to spare Baldur. It had been put into the blind god's hand in a friendly helpful sort of way by a designing female, who was really the evil spirit Loki in disguise. What is the allegory? Does the mistletoe dart stand for calumny? Is the mistletoe associated with calumny because it is a parasite in high places? If one must choose between the mistletoe myth of Norsemen and Briton—the latter, which survives in the power accorded to the mistletoe to license, even to command, by its mere overhead existence, the giving and taking of unexpected kisses and of expected ones, too, is certainly the more cheerful and suitable to the hopeful enterprise of New Year.

24. The Cattle Show

I always look upon the Christmas Cattle Show of the Smithfield Club as a scientific delight. Breeding is a most serious branch of scientific knowledge, held by many people (of whom I am one) to be of more importance to statesmen, politicians, and philanthropists than any other kind of knowledge, and yet almost absolutely neglected and completely ignored except by our farmers and horticulturists. When examining in turn the splendid animals at Islington I have felt indignant that it should be not improbable that, owing to ignorance and neglect in official quarters, the long matured traditions and built-up skill of our cattle-breeders will be destroyed, crushed out of existence by huge, devastating capitalist " combines." Soon we shall not get the beef we wish for, but we shall have to take whatever inferior stuff the giant monopolist chooses to force on us—or go without! Our wonderful stock, so patiently and happily bred, the envy of the world, will disappear, and our breeders forget their art. We shall none of us in Britain know more about prime beef, roasts, grills, and marrow-bones than do the people of Europe or the eaters of terrapin and soft-shelled crabs.

It is wonderful that man, by deliberate choice in selecting the sires and dams, has been able to produce such widely-different races as the short-horn, the High-land and the Sussex breed, and not only to produce them, but to keep them there generation after generation. In Nature, no such deviations are allowed—her motto is " One species, one shape," which is only relaxed so as to allow a few geographical varieties. It is man who makes all these strange breeds, just as he has made such a queer, irregular, varied lot of creatures from the human stock. Withdraw once and for all man's guiding " intelligence," or perversity, if you choose so to call it, and all these cattle would in a few hundred years revert to one form, nearly (but not quite) the same as that they came from. So, too, the Sheep ; so, too, the Pigs. And man himself, if one could poison him universally with a mind-destroying microbe, would become a beautiful, healthy, silly creature, dying at first by millions annually, and at last represented by a hundred thousand unvarying specimens, inhabiting the warm but healthy corners of the earth, aimlessly happy, free from disease, neither increasing nor decreasing in number. It is legitimate, and is a means of examining the whole problem of man's history, to inquire whether we have reason or not to suppose that, were intelligent man thus removed arbitrarily and completely from the scene, a new " lord of the world " would arise, by normal evolutionary process. A bird, an elephant, a rat, might give rise to the new line of progressive development, and, unchecked by man, once jealous and repressive, but now down-fallen, this new stock might acquire such brains and wits as we men now boast of, and people the earth. You never can tell ! But it is not the business of science to expatiate on such possibilities.

The domesticated cattle of Europe are of very ancient prehistoric origin. They are for convenience called " Bos taurus," and seem to be derived from the huge Bos primigenius or Aurochs, the Urus of Cæsar, which was wild in Central Europe in his time, and from the Indian Bos indicus—which is represented by the Indian and African native breeds of " humped " cattle. It is, however, very difficult to trace most of man's domesticated animals or his cultivated plants to their original wild forms and original habitation. At the Cattle Show we only see British and Irish breeds, and only those cattle bred as meat-makers—the Highland, the Welsh, the Shorthorns, the polled Angus, the South Devons, the Hereford, the Sussex, the Galloway, the Dexter. But there are other British breeds famous for their milk-producing quality, such as the Guernseys and Jerseys, whilst in Hungary, Italy, and Spain they have magnificent breeds of great size, and often with truly splendid spirally-turned horns (e.g. the Spanish), which are used for ploughing and carting, and are fattened, killed, and eaten after doing ten years' good work. These fine creatures are not seen in England. They come nearest to the extinct Aurochs, which was, however, bigger than any of them. It, too, existed in prehistoric times in England, and we find its bones in the gravel of the Thames Valley. The last aurochs, or wild bull of Europe, was killed in Poland near the end of the seventeenth century. The wild Chillingham cattle are Roman cattle run wild. Many of these breeds and the bones of the aurochs to compare as to size may be seen in the north hall of the Natural History Museum, where I commenced a collection of domesticated breeds of cattle, sheep, horses, dogs, &c., eight years ago. Chillingham cattle are to be seen in the Zoological Gardens.

An interesting fact in this connection is that the splendid bull which is kept in half-wild herds in Spain for the purpose of " bull-fights," is of a totally different race from that of the big, long-horned agricultural cattle. It may be seen at Cromwell-road, a specimen killed in the ring having been procured at my request and presented to the museum through the kindness of the British Consul at Seville. The Spanish fighting bull is, curiously enough, more like our Channel Island milk-producing cattle than any other. It probably came to Spain from North Africa—but there seems to be no record or history concerning it—and if there were it would probably be a fantastic invention. It seems that only the bulls of this special breed can be played with and dazzled by the matador's red cloak. A Scotch bull was once brought by sea to Seville and introduced to the arena. He paid no attention to cloaks, red or otherwise, but always went straight for his man. It is stated that he was soon left quite alone in the ring! The native African cattle (of Indian origin) at Ujiji and in Damaraland have the biggest horns of any true Bos—as much as $13\frac{1}{2}$ ft. along the curve from point to point. We have to distinguish from our own cattle, for which there is no name except "Bos taurus," for neither ox, bull, cow, heifer, nor steer will do—the other bovines—the buffaloes, the yak, and the bison—besides those great beasts the gayal and the gaur of India and the banting of Malay. All these may be seen and studied either in the Museum or the Zoological Gardens.

25. *The Experimental Method*

The observations lately made by a Chancellor of the Exchequer about an attempt to put salt on a bird's tail remind me of my first attempt to deal experimentally with a popular superstition. I was a very trustful little boy, and I had been assured by various grown-up friends that if you place salt on a bird's tail the bird becomes as it were transfixed and dazed, and that you can then pick it up and carry it off. On several occasions I carried a packet of salt into the London park where my sister and I were daily taken by our nurse. In vain I threw the salt at the sparrows. They always flew away, and I came to the conclusion that I had not succeeded in getting any salt or, at any rate, not enough on to the tail of any one of them.

Then I devised a great experiment. There was a sort of creek eight feet long and three feet broad at the west end of the ornamental water in St. James's Park. My sister attracted several ducks with offerings of bread into this creek, and I, standing near its entrance, with a huge paper bag of salt, trembled with excitement at the approaching success of my scheme. I poured quantities—whole ounces of salt—on to the tails of the doomed birds as they passed me on their way back from the creek to the open water. Their tails were covered with salt. But, to my surprise and horror, they did not stop! They gaily swam forward, shaking their feathers and uttering derisive "quacks." I was profoundly troubled and distressed. I had clearly proved one thing, namely, that my nursemaid, uncle, and several other trusted friends—but not, I am still glad to remember, my father—were either deliberate deceivers or themselves the victims of illusion. I was confirmed in my youthful wish to try whether things are as people say they are or not. Somewhat early perhaps, I adopted the motto of the Royal Society, "Nullius in verba." And a very good motto it is, too, in spite of the worthy Todhunter and other toiling pedagogues, who have declared that it is outrageous to encourage a youth to seek demonstration rather than accept the statement of his teacher, especially if the latter be a clergyman. My experiment was on closely similar lines to that made by the Royal Society on July 24, 1660—in regard to the alleged property of powdered rhinoceros horn—which was reputed to paralyse poisonous creatures such as snakes, scorpions, and spiders. We read in the journal-book, still preserved by the society, under this date: " A circle was made with powder of unicorne's horn, and a spider set in the middle of it, but it immediately ran out severall times repeated. The spider once made some stay upon the powder."

26. *Hypnotism and an Experiment on the Influence of the Magnet*

A more interesting result followed from an experiment made in the same spirit twenty-five years later. I was in Paris, and went with a medical friend to visit the celebrated physician Charcot, to whom at that time I was a stranger, at the Salpêtrière Hospital. He and his assistants were making very interesting experiments on hypnotism. Charcot allowed great latitude to the young doctors who worked with him. They initiated and carried through very wild "exploratory" experiments on this difficult subject. Charcot did not discourage them, but did not accept their results unless established by unassailable evidence, although his views were absurdly misrepresented by the newspapers and wondermongers of the day.

At this time there had been a revival of the ancient and fanciful doctrine of "metallic sympathies," which flourished a hundred years ago, and was even then but a revival of the strange fancies as to "sympathetic powders," which were brought before the Royal Society by Sir Kenelm Digby at one of its first meetings, in 1660. In the journal-book of the Royal Society of June 5 of that year, we read, " Magnetical cures were then discoursed of. Sir Gilbert Talbot promised to bring in what he knew of sympatheticall cures. Those that had any powder of sympathy were desired to bring some of it at the next meeting. Sir Kenelm Digby related that the calcined powder of toades reverberated, applyed in bagges upon the stomach of a pestiferate body, cures it by several applications." The belief in sympathetic powders and metals was a last survival of the mediæval doctrine of " signatures," itself a form of the fetish still practised by African witch-doctors, and directly connected with the universal system of magic and witchcraft of European as well as of more remote populations. To this day, such beliefs lie close beneath the thin crust of modern knowledge and civilisation, even in England, treasured in obscure tradition and ready to burst forth in grotesque revivals in all classes of society. The Royal Society put many of these reputed mechanisms of witchcraft and magic to the test, and by showing their failure to produce the effects attributed to them, helped greatly to cause witches, wizards, and their followers to draw in their horns and disappear. The germ, however, remained, and reappears in various forms to-day.

Thirty years ago some of the doctors in Paris believed that a small disc of gold, or copper, or of silver, laid flat on the arm could produce an absence of sensation in the arm, and that whilst one person could be thus affected by one metal another person would respond only to another metal, according to a supposed "sympathy" or special affinity of the nervous system for this or that metal. This astonishing doctrine was thought to be proved by certain experiments made with the curiously "nervous" (hysterical) women who frequent the Salpêtrière Hospital as out-patients. That the loss of sensation, which was real enough, was due to what is called "suggestion"—that is to say, a belief on the part of the patient that such would be the case, because the doctor said it would—and had nothing to do with one metal or another, was subsequently proved by making use of wooden discs in place of metallic ones, the patient being led to suppose that a disc of metal of the kind with which she believed herself "sympathetic" was being applied. Sensation disappeared just as readily as when a special metallic disc was used.

The old hypothesis of the influence of a magnet on the human body was at this time revived, and Charcot's pupils found that when a susceptible female patient held in the hand a bar of iron surrounded by a coil of copper wire leading to a chemical electric cell or battery nothing happened so long as the connection was broken. But as soon as the wire was connected so as to set up an electric current and to make the bar of iron into a magnet, the hand and arm (up to the shoulder) of the young woman holding the bar, lost all sensation. She was not allowed to see her hand and arm, and was apparently quite unconscious of the thrusting of large carpet-needles into, and even through, them, though as long as the bar of iron was not magnetised she shrunk from a pin-prick applied to the same part. I saw this experiment with Charcot and some others present, and I noticed that the order to an assistant to "make contact," that is to say, to convert the bar of iron into a magnet, was given very emphatically by Charcot, and that there was an attitude of expectation on the part of all present—which was followed by the demonstration by means of needle-pricking that the young woman's arm had lost sensation, or, as they say, " was in a state of anæsthesia."

Charcot went away saying he should repeat the experiment before some medical friends in an hour or two. In the meantime, being left alone in the laboratory with my companion as witness, I emptied the chemical fluid (potassium bichromate) from the electric battery and substituted pure water. It was now incapable of setting up an electric current and converting the bar into a magnet. When Charcot returned with his visitors, the patient was brought in, and the whole ritual repeated. There was no effect on sensation when the bar was held in the hand so long as the order to set the current going, and so magnetise the bar, had not been given. At last the word was given, " Make ! " and at once the patient's arm became anæsthetised, as earlier in the day. We ran large carpet-needles into the hand without the smallest evidence of the patient's knowledge. The order was given to break the current (that is, to cease magnetising the bar), and at once the young woman exhibited signs of discomfort, and remonstrated with Charcot for allowing such big needles to be thrust into her hand when she was devoid of sensation! My experiment had succeeded perfectly.

It would not have done to let Charcot, or anyone else (except my witness) know that when the order " Make " was given, there was no " making," but that the bar remained as before un-magnetised. The con-

viction of everyone, including Charcot himself, that the
bar became a magnet, and that loss of sensation would
follow, was a necessary condition of the "suggestion"
or control of the patient.   It was thus demonstrated
that the state of the iron bar as magnet or not magnet
had nothing to do with the result, but that the important thing was that the patient should believe that
the bar became a magnet, and that she should be in-
fluenced by her expectation, and that of all those
around her, that the bar, being now a magnet, sensation
would disappear from her arm.   With appropriate
apologies I explained to Charcot that the electric
battery had been emptied by me, and that no current
had been produced.   The assistants rushed to verify
the fact, and I was expecting that I should be frigidly
requested to take my leave, when my hand was grasped,
and my shoulder held by the great physician, who said,
"Mais que vous avez bien fait, mon cher Monsieur!"
I had many delightful hours with him in after years,
both at the Salpêtrière and in his beautiful old house
and garden in the Boulevard St. Germain.

There are few "subjects" in this country for the
student of hypnotism to equal the patients of the
Salpêtrière and other hospitals in France—and very
few amongst those who read, and even write, about
"occultism" and "super-normal phenomena" know the
leading facts which have been established in regard to
this important branch of psychology.   The study of the
natural history of the mind, its modes of activity, and
its defects and diseases is of fundamental importance
—but its results are often either unknown or greatly misunderstood by those who have most need of such knowledge, namely those who, mistaking the attitude of an
ignorant child for that of "a candid inquirer," try to
form a judgment as to the truth or untruth of stories of
ghosts, thought-transference, spirit-controls, crystal-
gazing, divining-rods, amulets, and the evil eye.

27. *Luminous Owls and Other Luminous Animals and Plants*

A correspondent lately described in a letter to a London newspaper what he believed to have been " a luminous owl," which was seen flying about at night in Norfolk. He mentioned the well-known fact that the dense greasy pateh of feathers on the breast of the heron is said to be luminous by many trustworthy observers. It is very probable that it was some carnivorous or fish-eating bird, which was thus seen in a luminous condition at night. The occurrence is much more in accordance with known facts than most people would suppose to be the case. Light, even strong light, is produced by many natural objects without the accompaniment of heat. We usually expect not merely fire where there is smoke, but heat—in fact, great heat, where there is light or flame. Yet there are many instances to the contrary, and the word " phosphorescence " is used to indicate a production of light without heat in reference to the fact that phosphorus is luminous, even when covered with water, although no appreciable heat accompanies the light such as we are accustomed to observe in ordinary "combustion" or burning.

There is more than one kind of phosphorescence. We separate the phosphorescence which is due to the oxidation of peculiar fatty matters in the bodies of plants and of animals (such as glow-worms) from that which is caused by the breaking or heating of crystals (white arsenic and apatite), or by longer or shorter exposure to the sun's rays (luminous paint), or by radio-activity, or by electrical discharges in vacuum tubes.

The "luminous owl" of the above-mentioned correspondent and the luminous breast of the heron probably

F

owe their strange appearance to the birds having smeared themselves with phosphorescent carrion or dead fish, the luminosity of which is due to bacteria.   The simplest case of phosphorescence in living things is that of the almost ubiquitous phosphorescent bacteria, minute microbes like those which cause putrefaction.   They can be obtained and cultivated from almost any sample of sea water.   A thin slice of meat placed in a shallow dish of salt water, so as to be barely covered by the liquid, will in cool, damp weather, almost certainly become covered with the growth of this phosphorescent germ and appear brilliantly luminous.   The populations of seaside towns have often been terrified by all the meat in the butchers' shops suddenly becoming thus phosphorescent.   The growth may be cultivated in flasks of salt broth.   I have prepared such flasks, which, when shaken so as to introduce oxygen, give out a heatless blaze of light of a greenish colour, brilliant enough to light up a room.   I once found a bone in a dog's kennel which was brilliantly phosphorescent owing to this bacterium.   I kept it for several days and showed it to Huxley as well as to other friends.   A certain kind of phosphorescent bacteria are parasitic in the blood of sandhoppers, causing a disease which kills them.   The diseased sandhoppers shine like glow-worms.   I have found them abundantly on the sea shore near Boulogne and near Trouville, but not yet on the English coast.   The bacteria can be seen with the microscope and inoculated from diseased luminous sandhoppers into healthy ones by using a needle to prick first the diseased and then the healthy creature.

The animals of the sea are often provided with secreting organs, producing a fatty body which can be oxidised and made luminous at the pleasure of the animal.   Thus many marine worms and minute sea-shrimps give out brilliant flashes of light.   Jellyfish of many kinds, and the minute noctiluca, no bigger than a pin's head, and the three-horned animalcule Ceratium tripos are the usual cause of the phosphorescence of the sea on our own coast. Deep-sea fishes are provided with large phosphorescent discs or plates on the surface of the body, which are sometimes furnished with lenses like a bull's-eye lantern. Glow-worms and fire-flies and some tropical beetles are examples of insects which have fatty phosphorescent organs which they can illuminate (oxidise) at pleasure, under the control of the nervous system. Some of the West Indian phosphorescent beetles are remarkable for having "lights" of two different colours. In the marshes around Mantua the fire-flies are so abundant at the end of June that the air for miles is full of them, and the sight so extraordinary and beautiful as to be worth a long journey to see. I have seen fire-flies as far north as Bonn on the Rhine. Once I was nearly upset by a horse shying at a glow-worm on a bank in Worcestershire. Some moulds and well-grown toadstools are phosphorescent, and a phosphorescent earthworm, a peculiar species, now well known, was first of all discovered in the South of Ireland by the late Professor Allman. In the autumn I have often picked up the phosphorescent centipede, which is remarkable for the fact that the phosphorescent material is a kind of slime which exudes from the body—the creature leaving thus a luminous trail behind it as it crawls. The piddock, or pholas—a boring sort of mussel—has brilliant phosphorescent glands, and the boys at Naples love to munch these shell-fish at night, and then to alarm the passer-by by opening their mouths, and showing a brilliant green light within. Cases are recorded, but not recently, of persons suffering from tuberculosis becoming phosphorescent; a possible, but certainly a rare, occurrence. Animal and vegetable phosphorescence is varied in colour. The light emitted is blue-green, green, yellow, orange, and even red in different cases. It is always due to the oxidation of a separate fatty chemical body, which can in many instances be extracted, then dried, and subsequently made luminous by moistening with ether, in consequence of which oxidation by the oxygen of the atmosphere is facilitated.

28. *Reminiscences of Lord Kelvin*

The late Lord Kelvin was one of the most fascinating personalities in the learned world. He uttered with a delightful simplicity the thoughts, however romantic and fanciful, which bubbled up in his wonderful brain. It was because he was so much of a poet that he was so great a man of science. Atoms and molecules and vortices, and the vibrations and gyrations of ether, and "sorting demons" were all pictured in his mind's eye, and used as counters of thought to give shape and the equivalent of tangible reality to his conceptions. By such conceptions he was able to present to himself and his listeners the complex mechanisms of crystals, of liquids, of gases, of electrical and magnetic currents, and the endless astounding proceedings of rays of light unsuspected by the ordinary man.

I think the last occasion on which he spoke in public was after Sir David Gill's brilliant address to the British Association at Leicester last August. Lord Kelvin was sitting close to me on that occasion, and I noticed that he never moved his gaze from the speaker. He followed Sir David's account of stars, whose distance is stated by the number of years it takes for their light to travel to this earth, like an enraptured schoolboy, and cheered when the evidence for the existence of two great streams of movement of the heavenly bodies, in opposite directions, going no one knows whither, coming no one knows whence, was sketched to us by the lecturer. In proposing a vote of thanks to Sir David Gill, Lord Kelvin burst into a sort of rhapsody, in which, with unaffected enthusiasm, he declared that we had been taken on a journey far more wonderful than that of Aladdin on the enchanted carpet; we had been carried to the remotest stars and well-nigh round the universe, and brought back safely to Leicester on the wings of science, and the most marvellous thing about it all was that it is true !

A few weeks before this Lord Kelvin was at the dinner in celebration of the jubilee of the foundation of the Chemical Society. In the speech which he then made he referred to the painful accident of a year or so ago which we had all so much regretted, when he had burnt his hand accidentally in some experiments with phosphorus, and had had to carry his arm in a sling for some weeks. "Lord Rayleigh, the president of the Royal Society," he said, "has just told us how, as a boy, he gave proof of his devotion to chemical science by burning his fingers with phosphorus—but I think my devotion must be considered greater than his, for I burnt my fingers very badly with phosphorus only last year, when I was 83 years old. It was at the end of April. My friends said I was old enough to know better, and it should have happened, not at the end of April, but on the first day, of that month." Lord Kelvin was associated in work in the sixties and seventies with another splendid man, Tait, of Edinburgh, who, besides being a great professor of "Natural Philosophy," and joint author of the celebrated treatise known as *Thomson and Tait*, was a great athlete—a golfer of the first class, a first-rate billiard player, and a wise lover of good ale, which he drank and gave to his friends to drink, whilst he discoursed as few, if any, to my knowledge, can now do, of things philosophical, mathematical, and humane.

29. *The So-called Jargon of Science*

It is often discussed as to whether science fails to obtain the attention of the public and to excite intelligent interest, owing to the obscure language which lecturers and writers use when attempting to expound scientific views and discoveries to " the ordinary man," or whether the fault lies with the " ordinary man " himself, who is too frivolous to bother about following carefully the words addressed to him, and, moreover, has never learnt even the A B C of science at school. It is certainly the case, as Professor Turner, the Oxford professor of astronomy, has pointed out, that a popular lecturer could tell his auditors a good deal more in an hour if they already had the elements of his subject at their fingers' ends than he can under the existing state of neglect of school education in the natural sciences.   That, however, seems to be obvious enough, and does not touch the real question.

I have had a long experience, both in lecturing myself and in assisting in the training of others to lecture and also to inform the uninstructed public by means of museum-labels and popular notes.   It seems to me that there are a large number of men who, even though capable of expressing themselves clearly under usual circumstances, yet fail to do so when trying to expound or to teach, in consequence of three distinct faults, any one of which is enough to render their discourse or writing hopelessly obscure to " the man in the street." These are, first, a kind of pride in using special terms and modes of expression which infatuates the lecturer or writer, and leads him, without reflection, to an attitude of mind expressed by saying, " That is the correct statement about this matter, short and true.   If you don't understand it, there are others who can.   You can leave it alone ; it is not worth my while to spend time and

trouble to explain further ; it is for you to give your-
selves the trouble to find out what I mean." The second
fault is a real incapacity (which occurs in many learned
men) to realise the state of mind of the uninstructed
man, woman or child who eagerly desires to be instructed :
this is want of imagination and want of sympathy.
There is no cure for those who fail as teachers for either
of these two reasons.

The third fault is much more widely at work, and the
most kindly sympathetic lecturers and writers—but more
especially lecturers—often suffer from it and could easily
amend their practice. It consists in the attempt to tell
the audience or reader too much—vastly too much—in
the limit of one hour, or within the space of a few lines
or pages. This failure is well-nigh universal. I have
heard a distinguished discoverer, an eloquent and able
man, try to tell a completely ignorant audience in one
hour the results of years of experiment and work by
many men on the electrical currents observed in nerves.
The audience did not know what is meant by an electrical
current, nor anything about nerves, nor a single one of
the technical terms necessarily used by the lecturer. The
task was an impossible one. In six lectures it might
have been accomplished, and great delight and increase
of understanding afforded to the listeners instead of
perplexity and a sense of their own incapacity and the
hopeless obscurity of science. That, I am convinced, is
the real trouble, viz., the attempt to tell too much in a
short time, the failure by the lecturer to arrange his
exposition in a series of well-considered, definite steps,
each exciting the desire to know more, and each given
sufficient time and experimental illustration or pictorial
demonstration to lodge its meaning and value safely and
soundly in the tender brain of the ignorant but willing
listener. I am convinced that there is in very many
lecturers a tendency to try to crowd and compress into
one lecture what should occupy ten—if the willing and

intelligent but ignorant listener is to feel happy and is really to understand what is said and done for his instruction. A special difficulty also arises from the fact that the lecturer often feels himself called upon to address and to say something to those among the audience who already know a good deal about his subject, as well as to make things clear to those who are absolute novices.

Some people have made this discussion the opportunity for attacking on the one hand the English language, and on the other the use of special names applied by men of science to special things and special processes. We cannot at once change the English language, even did we wish to do so. But the creation of special names to distinguish things not distinguished from one another in common speech is a necessity. It cannot be avoided. It is mere impatience and temper to call the names and terms which are necessary as counters of thought " jargon." No doubt there may be in some lecturers and writers a tendency to excessive use of special terms and names, but the real trouble in the matter arises from the too rapid thrusting of a large number of such unfamiliar words upon an untrained audience. If new words are introduced in moderation they can be assimilated. They cannot be dispensed. with altogether. A correspondent lately complained to me that I wrote of the minute creature which causes the sleeping sickness as a Trypanosome, whereas, had I called it " a blood-parasite " he would have known what I meant, and been able to follow my statement more easily. I am sorry to say that I cannot agree with him. There are many kinds of blood-parasites; there are the worms known as Filariæ, there are the vegetable microbes known as bacteria and bacilli and spirilla, and there are minute creatures of an animal nature called pyroplasma and trypanosoma (beside some others). These must be distinguished from one another if we are to understand anything about the causation of disease by microbes. It would be mere muddling and confusion to simply call them all by the same name, simply " blood parasite." That would cause the same sort of confusion as would occur if the Smiths or Browns of our acquaintance had no Christian names by which we can separate each member of the class from the others and assign to him his own special qualities, opinions, and property. What some people call " scientific jargon " is assuredly not a thing to be proud of or to mouth with a sense of superiority. Nevertheless, it is absolutely necessary, and must be introduced gently and considerately to the stranger who can and will, if reasonably handled, appreciate the immeasurable advantage of having distinct words to signify distinct things. That, after all, is an elementary feature in all language. And just as the " jargon " of a game, a sport, or a profession has a fascination for those who use it, and forms a bond of union or special understanding between them, so inevitably does the jargon of a branch of science flourish in the thought and on the lips of those who devote themselves to that branch, and bind them in a sort of freemasonry. We do not expect cricketers or golfers to talk in plain English ; why should we expect chemists or naturalists to do so ? After all, it is a question of moderation and of gradually increasing the dose. The beginner must not be terrified by an array of outlandish words.

30. *Rats and the Plague*

Rats ! Who said rats ? That is an important question, because the word means different things to different people. To some persons " rats " means simply " nonsense " ! To Sir James Crichton Browne it means the devastator of stores and the dread carrier of bubonic plague. To the naturalist it means a group or natural cohort of small mammals similar to our common rat and mouse, representatives of which are found in every quarter of the globe and in almost every island of the sea. The distinct "kinds" or "species" are numbered by the hundred. They are extraordinarily alike, and can only be distinguished and classified into proper "species" by careful examination and measurement. Mr. Oldfield Thomas, of the Natural History Museum, has made a special study of them. To give an idea of his work, it may be mentioned that ninety different names had been given by previous writers to as many apparently distinct kinds of rat occurring in India. But by careful measurement and study of the relations to one another of these rats, Mr. Thomas has reduced the number of really distinct Indian species of rats and mice (for a mouse is only a smaller rat) to nineteen. What we call in English water-rats, or water-voles, field-voles, and such little foreign beasts as the lemming and the hamster, are very close to rats in appearance, but are separated on account of clear differences of structure from true rats and mice.

At a meeting in London the total destruction of "rats" was advocated. Whether it was affirmed at the meeting, or was merely an error of those who wrote and commented on the matter afterwards, I do not know, but it was very generally stated in this connection that the old Black rat (known to naturalists as Mus rattus) is quite extinct in England, and that its place has been taken by the Norwegian, or Grey rat (Mus decumanus), also called the Hanoverian rat, because it became noticeable by its abundance in this country at the time of the accession of the Hanoverian kings. The Black rat is not extinct in England, not even very rare. Mr. Stendall lately sent me specimens caught in his warehouse in the City of London, where they are abundant.

In many localities, *e.g.* Great Yarmouth, and in isolated dwelling-places they occur, and even outnumber the Norwegian rat. A most important and remarkable fact is that the rats which infest ships are often all Black rats. The Black rat, or Alexandrine rat (as Mr. Thomas calls it), lives in our houses, in the roof, in recesses of woodwork. It is a house rat, whereas the Grey, or Norwegian rat, lives in the sewers and the banks of ditches, and only comes up into the basement of houses through defective building. The Grey rat has driven out the water-voles from many river banks near towns, just as he has to a great extent taken the place of the Black rat in houses where the kitchen and food stores are close to and in communication with the sewer!

The Black rat cannot be really distinguished by his blackness. That is why some naturalists call him the Alexandrine rat, so as to avoid a misleading implication. He is often of a bright yellowish-brown colour along the back—with longer dark-brown hairs and a good deal of grey elsewhere—quite like the Norwegian or Grey rat in colour. At the same time he is often blackish, and frequently very black. The colour of all these kinds of rats and mice can vary, according to the conditions and colour surroundings in which they live. Black, white, sandy-brown, or a mixture of spots of all three colours, or a uniform " mouse-brown " tint, are (as most boys know) the possibilities revealed by allowing them to breed in captivity. Nature selects accordingly the particular tint which affords protection from observation by enemies in a given locality.

The real distinction between the Black (Alexandrine) rat and the Grey (Norwegian) rat is that the Black rat is smaller, has a tail longer than its body (125 per cent.), and long and wide ears, which stand out from the head. The Grey (Norwegian) rat is a larger, heavy-bodied rat, with a tail shorter than its body (90 per cent.), and short ears. Both these rats are common in India, but there is a third kind, which is the commonest of the three in Calcutta, and is probably the one most concerned in the dissemination of plague. It differs in some definite features from both the Black rat and the Grey rat, although it is very much like the latter in general appearance. It is called Nesokia Bengalensis, or Mole-rat. It is a big rat—its tail is only 70 per cent. the length of its body; the pads on the soles of its feet differ from those of the two other rats; its fur is thin and bristly, and when it is put into a cage it erects its bristles and spits! It is, like the Black rat, a stable and granary rat, and makes burrows in which it stores grain.

The rats of Calcutta have been carefully studied lately by Dr. Hossack, in consequence of their connection with the bubonic plague. In the older native parts of Calcutta, the Mole rat is twice as common as the Norwegian Grey rat, and the Black rat not so abundant as the latter. In the central European part of the town the Grey rat is commoner than the Mole rat—because, apparently, the better-built houses do not afford such facilities for burrowing. The Black rat is here also by a good deal the most uncommon of the three. All these rats suffer from the plague, die from it, and the fleas which lived in their fur leave them as they get cold, and make their way on to human beings, whom they consequently infect with the plague bacillus. This has now been quite conclusively proved by the Indian doctors charged by Government with the study of the causes of the plague. The plague bacillus —a minute, rod-like organism, which grows in the blood and lymph, once it has effected a lodgment, and there produces deadly poison—was discovered some fourteen years ago, but it is only recently that the plague bacillus has been shown to live in the intestine of the flea, which sucks it up with the blood or other fluids of the rat on which it lives. The flea, which readily goes to man, does not suffer from the plague bacilli which it has gorged, but conveys them to man either by its bite or by its excrement.

This being so, it becomes important to know all about the fleas of rats. Quite unexpected facts have been discovered in regard to them. In Europe a very large flea is found on the grey and the black rat. This kind has not, I believe, ever been found on human beings or been known to bite them. But in India, in the Philippines, and in the ports of the Mediterranean, this northern rat-flea is rare, and its place is taken by a smaller and more actively vagrant flea, which Mr. Charles Rothschild (who is the great authority on fleas) found upon several different kinds of small animals in Egypt. IIe named it "Pulex cheopis." This is the flea (and not our big northern rat-flea) which acts as the carrier of plague-germs from rats to man in India. It appears from experiments that the common flea of man (Pulex irritans) and the cat-and-dog flea (Pulex felis), as well as the big northern rat-flea (Ceratophyllus fasciatus), can harbour the plague-bacillus if fed on plague-stricken animals, but there are no observations to show (as there are about the "Cheops flea") that they pass habitually from man to rats and rats to men.

It is happily so long (200 years) since we had a real outbreak of plague in Europe that we are still in doubt as to whether the Grey rat or the Black rat is the more susceptible to the disease—and what flea, if any, acts, or has acted, as the carrier from rat to man in this part of the world. The suggestion has been made that the Grey Norwegian rat takes plague less easily than the Black rat, or than the Indian Mole-rat (Nesokia), and that the multiplication of the Grey rat in England and France and consequent decrease in Black rats, is, therefore, an advantage, so far as plague is concerned. Possibly with the Grey rat has come the big rat-flea.

which does not attack man as does the Cheops flea. The disappearance of plague in Western Europe seems to correspond in date with the arrival of the Grey rat. But, on the other hand, an alteration in the character of our houses and their greater " accommodation " for the new rat rather than the old black species may account both for the increase of the latter and for the absence of dirt and vermin in the dwelling-rooms and bed-chambers which formerly enabled the plague-bacillus to flourish amongst us, and to reach the human population—as it does now in India and China. All this shows how necessary it is to have accurate true knowledge of such despised creatures as rats and fleas, if we are to live in great crowded cities closely packed together. And it should also make us try to gain further knowledge as to these creatures, so that we may form a reasonable anticipation of the consequences we are bringing down on our heads when we set about exterminating this or that race of animals. We are not yet sure that the Norwegian Grey rat is not a blessing in disguise.

31. *Ancient Temples and Astronomy*

Janssen, the French astronomer, who died about the same time as Lord Kelvin, acquired celebrity by his discovery of a method for seeing and studying the great flames or prominences which surround the sun. The glare of the great fiery ball is such that the eye is blinded in ordinary circumstances to the light of these prominences. They were only known from their coming into view during the total eclipse of the sun's disc by the moon. Then they were seen as a great fringe of pointed, tongue-like flames around the darkened disc. But at other times no use of smoked glass or telescope could bring them into view. Janssen went to India in 1868 to study these prominences of the sun during the total eclipse of that year. His purpose was to examine with a spectroscope the light given out by the prominences. The day after the eclipse Janssen found that he could still examine the prominences and make out their shape and the chemical elements present in them by looking at them through the spectroscope, although the sun's disc was now uncovered, and it was impossible to see the prominences with the unaided eye or with the telescope.

A young English astronomer, hundreds of miles apart from Janssen, on the same day, Aug. 18, 1868, made the same discovery in the same way, independently. The English astronomer was Norman Lockyer, and the French Academy of Sciences caused a medal to be struck in commemoration of this discovery. The medal is before me as I write. It shows the heads of Janssen and of Lockyer side by side, as they were forty years ago.

Each has carried on his researches and discoveries with unabated vigour since that happy conjunction. Sir Norman Lockyer has for many years added to his constant study of the sun, fixed stars, and nebulæ by means of the spectroscope and photographic record of spectra, an inquiry into the evidence afforded by astronomical facts first as to the age of Greek and Egyptian temples, and latterly as to that of the mysterious avenues and circles of stones (such as Stonehenge) scattered about the British Islands, of the history and use of which we have only vague traditions and no actual records. These stone circles and avenues are very numerous in Great Britain. The chief are Stonehenge, Avebury, and Stanton Drew in the middle South of England; the Hurlers, Boscawen-Un, Tregaseal, the Merry Maidens, and the Nine Maidens in Cornwall; Merrivale Avenue and Fernworthy Avenue in Devon; many circles in Aberdeenshire, in Cumberland,

Derbyshire, and Oxfordshire, as well as monuments of the same kind in Wales. Sir Norman Lockyer has obtained measurements of most of these and plans showing the relations of the principal lines of their ground plan to the points of the compass, and so to the position occupied by the sun and by certain stars on given days of the year at the rising or setting of those heavenly bodies. It may well be asked what is Sir Norman's object in doing this?

The explanation is as follows: The builders of Christian churches in Europe have, as a rule, set out the ground plan of the church shaped like a Latin cross, so that the arms of the cross run north and south —the head points to the east, or Orient, and the base to the west. In consequence of this custom the word " orientation " has come into use, to signify the direction purposely given to the main length of a temple or church. Now it appears that many, if not all, ancient temples (including the ancient stone circles and avenues of Britain) were purposely so " oriented " by their builders that a particular star, or the sun itself, should at a fixed day and hour in the year be seen during its movement across the heavens through an opening in the building especially designed for this purpose, so as to allow the light of the star to fall into the most sacred part of the temple, the " Naon," or Holy of Holies. At the moment of its appearance special ceremonies were performed by the priests and worshippers in the temple. The temple was dedicated to and carefully " oriented to" that particular star. Thus, in ancient Greece, the Pleiades, Sirius (the dog star), Spica, and other stars were thus used; in Egypt, Capella, Canopus, and Alpha Centauri; in Britain, Arcturus, as well as those used by the Greeks.

These temples were really astronomical observatories, and were meant always to remain " oriented " to their special star, which must, if the earth were steady in its position, although spinning like a top, and also circling round the sun, duly appear each year at the expected day and minute in the special "window" or aperture designed so as to allow the star—then, and then only—to shine into the temple. But the astronomers have discovered that the earth is not steady ! It "wobbles" very slowly and regularly as a top wobbles. The position of the axis of rotation—corresponding in position to the stem of a top—does not remain one and the same, but is pulled aside by the attraction of the sun and moon, and moves round as one may often see in the spinning of a top. The earth takes about 26,000 years for its poles to complete the cycle of its wobble. Moreover, in addition to this, there is the fact that the earth's axis (stem of the top) is not nearly upright, but inclined at a considerable angle (23 deg.) to the horizontal or plane of its orbit round the sun, and that this inclination very slowly changes, in addition to the wobbling movement. The amount and rate of these changes in the inclination of the axis of the earth have been definitely ascertained by astronomers.

I mention the nature of these movements because they clearly enough must upset altogether the desired result of the orientation of temples. The last-mentioned slow increase of obliquity affects solar temples chiefly, and the more rapid wobbling affects the star temples—both to such a degree that temples oriented two or three thousand years ago are now quite out of line, and no longer "catch," so to speak, their particular star or the sun on the appointed day. They no longer point truly, because the "pitch" of the earth has altered since they were set.

The next point is that astronomers are able to calculate with surprising accuracy from other observations how much exactly at this moment the "pointing," or "alignment," must be "out" as compared with a thousand, fifteen hundred, two, three, four, or more thousand years ago. Accordingly, if you know the star to which an ancient temple was set or aligned, the day of the solar year which was the festival or critical moment of the appearance of the star in the sacred aperture—and how much the temple is to-day out in its pointing, that is to say, the exact amount of swinging which would bring the temple back into its original relation to the star—you have a means of measuring the age of the temple; you have a measure of the time which has elapsed since it acquired this amount of departure from correct orientation. Astronomy tells you how much it must get out of line in every hundred years.

Mr. F. C. Penrose, F.R.S., investigated this matter in regard to several Greek temples; others besides Sir Norman Lockyer have written on the aberration and calculable age of Egyptian temples. It has, for instance, actually been found that the temple of Ptah was aligned to the sun in the year 5200 B.C. The alignment is no longer correct, and it appears that the Egyptians themselves discovered that some of their most ancient temples had lost correct alignment, and erected new and corrected buildings in connection with them, and re-dedicated them. Now Sir Norman is making a vigorous effort to procure all the possible measurements and indications concerning the prehistoric circles and avenues of Britain before it is too late. They are being more and more rapidly destroyed. Stonehenge has been carefully measured and its present alignment determined by various surveyors. Its age is discussed by Sir Norman Lockyer in an interesting book, but we may soon expect a further discussion of the whole subject of these prehistoric British monuments from his pen. In some cases, as in that of Stonehenge, the relation of the temple to the sun is obvious and confirmed by tradition and existing custom. But in many cases investigation is rendered very difficult by the absence of any immediate indication of what precisely is the heavenly body to which the temple was at its foundation oriented.

In the case of Stonehenge, the conclusion at which Sir Norman Lockyer arrives is that there was an earlier circle of small stones (still represented), but that the temple was rededicated, and the larger trilithons (each consisting of two uprights and a cross-piece) erected, and the main opening of the circle aligned to the midsummer rising sun about 1700 B.C., with a possible error of 200 years, more or less. This is arrived at by measurements showing the exact amount by which the alignment is "out" at the present day. This date is confirmed by the recent discovery of numerous stone hammers when one of the big stones was dug under and restored to the upright position from which it had slipped. The stone age is believed to have given place in Britain to the use of metal before 1700 B.C., and no metal tools were found at Stonehenge.

Stonehenge—the most wonderful, mysterious, and complete of the great astronomical temples of Western Europe—has come down to us from the absolute darkness of prehistoric ages. Its secrets are still buried in the ground around and under its huge monoliths. This prodigious relic of the past is actually the private possession of one happy man, Sir Edmund Antrobus. Only two years ago he earned the gratitude of all men by employing workmen and machinery, at considerable expense, to restore one of the great stones to its upright position. The extraordinary thing is that whatever money is needed for the purpose is not at once offered to enable him to examine and replace with scrupulous care every stone, big and small, every scrap of soil, within an area of many hundred yards, embracing Stonehenge and all around it. I understand that he is willing to sell this great possession to the nation. It surely ought to be acquired as national property, and reverently excavated and preserved, whilst every fragment of significance found in the excavations should be placed in a special museum at Amesbury or Salisbury, under unassailable guardianship. Year by year it has crumbled away. We owe the sincerest thanks to Sir Edmund Antrobus for having placed a light wire fence around the venerated relics, and for putting a guardian in charge so as to arrest, even at this latest moment, the final desecration and destruction of this splendid thing by heedless ruffians. The protection afforded is, nevertheless, insufficient. The delay in examining everything on the spot and in making all that remains absolutely secure is a national disgrace.

32. *Alchemists of To-day and Yesterday*

The claim to have devised a secret process in virtue of which sugar or charcoal placed in an iron crucible and heated to a tremendous temperature is found on subsequent cooling to contain large marketable diamonds has a close similarity to the pretensions of the alchemists. It differs in the fact that very minute diamonds have actually been formed by a scientific chemist (M. Moissan) in such a way, whilst the alchemists' search was for a substance—the " philosopher's stone," as it was called, which was never discovered, but was supposed to have the property, if mixed and heated in a crucible with a base metal, of converting the latter into gold. From time to time those engaged in this search honestly thought that they had succeeded ; others were impostors, and others laboured year after year, led on by elusive results and dazzling possibilities.

In England, after the true scientific spirit had been brought to bear on such inquiries by Robert Boyle and the founders of the Royal Society in the later years of the seventeenth century, little was heard of "alchemy," and the word "chemistry" took its place, signifying a new method of study in which the actual properties of bodies, their combinations and decompositions, were carefully ascertained and recorded without any pre-possessions as to either the mythical philosopher's stone or the elixir of life. But as late as 1783—only a hundred and twenty-five years ago—we come across a strange and tragic history in the records of the Royal Society associated with the name of James Price, who was a gentleman commoner of Magdalen Hall, Oxford. After graduating as M.A., in 1777 he was, at the age of twenty-nine, elected a Fellow of the Royal Society of London. In the following year the University of Oxford conferred on him the degree of M.D. in recognition of his discoveries in natural science, and especially for his chemical labours. Price was born in London in 1752, and his name was originally Higginbotham, but he changed it on receiving a fortune from a relative.

This fortunate young man, whose abilities and character impressed and interested the learned men of the day, provided himself with a laboratory at his country house at Stoke, near Guildford. Here he carried on his researches, and the year after that in which honours were conferred on him by his university and the great scientific society in London, he invited a number of noblemen and gentlemen to his laboratory to witness the performance of seven experiments, similar to those of the alchemists—namely, the trans-mutation of baser metals into silver and into gold. The Lords Onslow, Palmerston, and King of that date were amongst the company. Price produced a white powder, which he declared to be capable of converting fifty times its own weight of mercury into silver, and a red powder, which, he said, was capable of converting sixty times its own weight of mercury into gold. The preparation of these powders was a secret, and it was the discovery of them for which Price claimed attention. The experiments were made. In seven successive trials the powders were mixed in a crucible with mercury, first four crucibles, with weighed quantities of the white powder, and then three other crucibles with weighed quantities of the red powder. Silver and gold appeared in the crucibles after heating in a furnace, as predicted by Price. The precious metal produced was examined by assayers and pronounced genuine. Specimens of the gold were exhibited to his Majesty King George III., and Price published a pamphlet entitled " An Account of Some Experiments, &c.," in which he repudiated the doctrine of the philosopher's stone, but claimed that he had, by laborious experiment, discovered how to prepare these composite powders, which were the practical realisation of that long-sought marvel. He did not, however, reveal the secret of their preparation. The greatest excitement was caused by this publication appearing under the name of James Price, M.D. (Oxon.), F.R.S. It was translated into foreign languages, and caused a tremendous commotion in the scientific world.

Some of the older Fellows of the Royal Society, friends of Price, now urged him privately to make known his mode of preparing the powders, and pointed out the propriety of his bringing his discovery before the society. But this Price refused to do. To one of his friends he wrote that he feared he might have been deceived by the dealers who had sold mercury to him, and that apparently it already contained gold. He was urged by two leading Fellows of the society to repeat his experiments in their presence, and he thereupon wrote that the powders were exhausted, and that the expense of making more was too great for him to bear, whilst the labour involved had already affected his health, and he feared to submit it to a further strain. The Royal Society now interfered, and the president (Sir Joseph Banks) and officers insisted that, " for the honour of the society," he must repeat the experiments before delegates of the society, and show that his statements were truthful and his experiments without fraud.

Under this pressure the unhappy Dr. Price consented to repeat the experiments. He undertook to prepare in six weeks ten powders similar to those which he had used in his public demonstration. He appears to have been in a desperate state of mind, knowing that he could not expect to deceive the experts of the society. He hastily studied the works of some of the German alchemists as a forlorn hope, trusting that he might chance upon a successful method in their writings. He also prepared a bottle of laurel water, a deadly poison. Three Fellows of the Royal Society came on the appointed day, in August, 1783, to the laboratory, near Guildford. It is related (I hope it is not true) that one of them visited the laboratory the day before the trial, and, having obtained entrance by bribing the housekeeper in Price's absence, discovered that his crucibles had false bottoms and recesses in which gold or silver could be hidden before the quicksilver and powder were introduced. Dr. Price appears to have received his visitors, but whether he commenced the test experiments in their presence or not does not appear. When they were solemnly assembled in the laboratory he quietly drank a tumblerful of the laurel water (hydrocyanic acid), which he had prepared, and fell dead before them. He left a fortune of £12,000 in the Funds. It has been discussed whether Dr. Price was a madman or an impostor. Probably vanity led him on to the course of deception which ended in this tragic way. He could not bring himself to confess failure or deception, nor to abscond. He ended his trouble by suicide. He was only thirty-one years of age! Not inappropriately he has been called the " Last of the Alchemists," though a long interval of time separates him from the last but one and the days when the old traditions of the Arabians' al-chemy were really treasured and the mystic art still practised.

33. *A Story of Sham Diamonds and Pearls*

It has been recently declared by a dealer in precious stones that though diamonds and other stones can be very well imitated, yet pearls cannot be.   This is hardly correct, as artificial pearls so well made as to defy detection by the casual glance of any but a professional expert are common enough.   Who does not know the pathetic story by the greatest of French writers, Guy de Maupassant, of the wife of a poor Government clerk, who borrowed a necklace from another lady to wear at a reception at the " Ministry " ? She lost the necklace (I forget whether it was of pearls or of diamonds, or both); but she and her husband were too proud to confess the fact, and purchased another necklace exactly like the lost one, for a sum the outlay of which reduced them for the rest of their lives to a state of penury and social exile.   They returned the new necklace in place of the lost one without a word, and accepted their fate.   By chance, the poor ruined lady, fifteen years afterwards, met her old friend, who had long since passed from her ac- quaintance, together with other prosperous people. Moved by her former friend's kind reception, she related the true history of the pearl necklace of long ago.   " Great   heavens ! "   exclaimed   the prosperous lady.   " The   necklace   I   lent   you   was   made   with imitation gems !   It was not worth five pounds ! " Too late !   Nothing now could give back to the high-minded, self-respecting little couple the lost years of youth passed in privation and bitterness.

34. *The Nature of Pearls*

Pearls have been lately studied by zoologists, and their true history made known. They are a disease, caused, like so many other diseases, by an infecting parasite. It is common knowledge that they are found much as we see them in jewellery, as little lustrous spheres embedded in the soft bodies of various shellfish, such as mussels, oysters, and even some kinds of whelks. They are not found in the shellfish like crabs and lobsters, called Crustacea, but only in those like snails, clams and oysters, called Mollusca. Pink pearls are found in some kinds of pink-shelled whelks. A pearl-mussel or pearl-oyster has a pearly lining to its shell, which is always being laid down layer by layer by the surface of the mussel's or oyster's body, where it rests in contact with the shell, which consequently increases in thickness. If a grain of sand or a little fish gets in between the shell and the soft body of its maker, it rapidly is coated over with a layer of pearl, and so a pearly boss or lump is produced, projecting on the inner face of the shell, and forming part of it. These are called " blister-pearls," and are very beautiful, though of little value, since they are not complete all round, but merely knobs of the general "mother-of-pearl" surface. These blister-pearls can be produced artificially by introducing a hard body between the shell and the living oyster or mussel.

It used to be thought that the true spherical pearls were caused by a hard granule of some kind pressing its way into the soft substance of the shell-fish, pushing a layer of the pearl-producing surface like a pocket in front of it. But it is now known that this "pushing in" is the work, not of an inanimate granule, but of a minute parasitic worm, which becomes thus enclosed by a pocket of the outer skin. The pocket closes up at its neck, and lays down layer after layer of pearl substance around the intrusive parasite, the dead remains of which can be detected with the microscope in sections of the pearl forming there a central kernel or nucleus. These parasitic worms were first detected in the small pearls formed by the common edible sea-mussel.

Though they are very small, sea-mussel pearls are collected for the market at Conway, in North Wales, and also on the coast of France. The parasitic worm is the young of a worm which, when adult, lives in the intestine of carnivorous fishes. It appears that it has to pass from and with the mussel into shellfish-eating sea fishes, where, although the mussel is digested, the parasite is not, but grows in size and alters its shape considerably. Then after a time the worm is swallowed, with the fish in which it has fixed itself, by sharks, dog-fish, and such fish-eating fishes. In these at last it becomes adult and of some size, an inch or so long, varying according to the particular kind, and produces many thousands of eggs, which hatch out as minute creatures swimming in the sea-water, and fortunate if they fall upon a bed of mussels. They enter the mussel's shell and make their way into its soft substance. A certain number (very few) get encased in the skin and covered up by pearl-layers, which is the mussel's way of killing them and putting them out of mischief. The others which have entered other regions of the mussel's body thrive, and have a chance of being swallowed by a mussel-eating fish, and then a further chance of that fish being eaten by a shark. If this happens the lucky worm—like the Italian who gets a winning number in three successive drawings of a lottery—gains the big prize. He becomes adult and produces innumerable young, who in their turn enter upon the chanceful career of a mussel parasite.

Thus we see that a pearl is not only a disease or abnormal growth caused by a parasite, but is actually an elaborately formed tomb or sarcophagus, in which the parasite is enclosed layer upon layer. This mode of disposing of parasites and other intrusive bodies is not unusual in animals. The terrible little flesh-worm—the Trichina—which causes the death of rats, pigs, and men who eat raw meat, is sometimes conquered in this way. It is found in the muscles (flesh) of man and animals enclosed in little pearl-like sacs, half the size of a hempseed, and it dies there, unless the invaded animal should die, and its flesh be eaten (as raw ham for instance) by another animal. The burying of inconvenient corpses in plaster of paris, corresponding to pearls as we now know them, has been a method of concealment occasionally adopted by criminals. On the whole, pearls have not very pleasant associations.

The history of the special parasitic worm which invades the beautiful little pearl-oyster of Ceylon has recently been followed out by skilful naturalists. There, too, a smaller oyster-eating fish of a peculiar kind, and a larger fish which eats the first fish, are necessary for the reproduction and multiplication of the pearl-producing parasites. The new Ceylon Pearl-Fishing Company has, therefore, to see to it that both these kinds of fish are encouraged to live in the sea near where the pearl oysters are found, and it is their object to increase the parasitic disease by which pearls are formed, and ensure an abundance of parasites.

An interesting new method has been recently applied to the examination of pearl oysters for pearls. The Rontgen rays are used to produce a skiagraph (such as surgeons use in searching for a bullet) of the pearl oysters when brought into harbour. They are thus rapidly examined one by one, without injury, and the shadow-picture shows the pearl or pearls inside those oysters which are infected. The pearlless oysters are returned to the depths of the sea, whence they came—those with small pearls only are kept in special reserves or sea-lakes, in order that the pearl may grow in size, whilst only those with good-sized pearls are opened at once, in order that the pearl may be extracted and sent to market.

There were great findings of pearls in the fresh-water pearl mussels of the Scotch rivers in former days. In the last forty years of the eighteenth century these pearls were exported from Scotland to France to the value of £100,000.

In the eighteenth century not only did they get their pearls from European rivers instead of from the East; but, instead of being excited about the artificial production of diamonds, they were driven wild with astonishment by the demonstration of the volatilisation of these stones—the disappearance of diamonds into invisible vapour when sufficiently heated. That the hardest stone in nature could be thus dissipated into thin air seemed incredible. On Aug. 10, 1771, a chemist named Rouelle invited to his laboratory to witness this wonder a company comprising the Margrave of Baden and the Princess his wife, the Dukes of Chaulne and of Nivernois, the Marchionesses of Nesle and of Pons, the Countess of Polignac, and some members of the Academy of Sciences, including the great chemist Lavoisier. Four diamonds— the largest belonging to the Count Lauraguais—were submitted before the eyes of all to the heat of a furnace, and in three hours had completely evaporated. There was, no doubt, room here for a mystification and for the abstraction of the diamonds with a view to dishonest appropriation. But no such purpose existed. The experiment was a genuine one, and Rouelle and his brother were honest investigators. They established the fact, now demonstrated as a lecture experiment, that the diamond is volatilised at very high temperatures. A more celebrated "evaporation" of diamonds—that which is known as "the affair of the Queen's necklace"

—took place a few years later in Paris, when no scientific investigation was connected with the embarrassing disappearance of the Royal trinket.

35. *A King Who was a Zoologist*

The King of Portugal, Carlos di Braganza, who was assassinated in the spring of 1908, was one of the most gifted and vigorous men of his age, fearless and intelligent to a rare degree, good-hearted, and devoted to the welfare of his people. If any man were justified in having no fear of outrage because he was conscious that his uprightness was proved and known to all men, his benevolence experienced by all, his ability and vast knowledge recognised by all, Dom Carlos was that man. Fanaticism, however, takes no account of the virtues of its victims. Until society has invented a method for keeping instruments of destruction out of the reach of dangerous, more or less maniacal individuals, all those who excite the fanatic's brain, even by the excellence and nobility of their lives, risk death whenever they trust themselves to the tender mercies of a crowd. Psychology may one day enable us to detect, and improved supervision of children enable us to segregate before it is too late, the latent assassins in our midst. If they have not a king as their quarry their reason is palsied by a president, and were there no presidents, they would become homicidal in the presence of a prefect or a policeman—even of a professor.

Some four years ago I had the honour of conducting Dom Carlos round the Natural History Museum in Cromwell Road. He arrived without attendant or escort, and I passed two hours alone with him. I had been told that he was a great shot and fond of natural history, that he played every athletic game, rode, and swam better than the best, that he was a fine water-colour painter, a real artist—and a first-rate musician and singer. I was astonished at his knowledge and personal experience in natural history. His burly form and bright, honest face gave me a most agreeable impression, and when he said (as I had been told he would) to each explanation of a specimen upon which I ventured for his edification, " I know! I know!" I felt that it was true, and that he really did know. " I have shot thirty of them in the south of my country," he said of some rare bird. " I know! I know! I have described a new species like that in my book on the birds of Portugal. I shall send it to you!" was his comment on another. When we came to some wonderful coral-like specimens—sea-pens and sea-feathers, dredged in the deep sea and preserved in spirits, for exhibition in the Museum—he said, to my astonishment, " Those are very bad. I get much better than those in my yacht off the Portuguese coast. I preserve them myself; it is a real art. I shall send you some." I said they would be a very welcome addition. " Yes, I know! I know!" he said. " Would you like some fishes, too? The Prince of Monaco has some fine things, and he led me to collect also myself. I have now many things better than his. I shall send you some fishes, too." And he did. A few months after his return to Portugal he sent to the Museum a large collection, preserved in spirit, which included many very fine and interesting specimens of deep-water Atlantic fishes; also his work, with coloured plates, on the Birds of Portugal, and a most remarkable publication on the tunny fisheries of the South Coast of Portugal—giving a careful survey of the waters, sea bottom, currents, fauna, and flora in correct, expert form, such as might issue from a Government Fisheries Board, but in this case done, as modestly indicated on the title-page, by the Head of the State himself, " Dom Carlos di Braganza." He went into the work-rooms of the Museum, where some new fishes were being drawn, and conversed with the naturalist in charge, and criticised the drawings. He saw everything, appreciated everything, and then looking at his watch, said, " I have only five minutes to get to a lunch party. Thank you very much for the most delightful time. I should like to stay all the day ; it is a splendid place," and was off in his brougham.

I exhibited the specimens and books sent by his Majesty for some weeks in the Central Hall of the museum, before they were incorporated in the great collection, for I felt that it was a rare and interesting thing that a king should not merely take a sportsman's pleasure in birds, beasts, and fishes, but actually be, so to speak, " one of us "—a zoologist who discovers, describes, and names new things. The Prince of Monaco is the only other head of a State who is a serious scientific naturalist. He has built and endowed a magnificent museum and laboratory at Monaco, where his skilled assistants carry on researches and look after the extremely valuable and important collections which he has himself made in a series of cruises in the Atlantic extending over many years. He has not only employed capable naturalists to help him, but is himself the chief authority and an original discoverer in " oceanography," the science of the great oceans.

A year or so ago, when Dom Carlos visited Paris, a special fête and reception was organised in his honour at the " Muséum d'Histoire Naturelle," in the Jardin des Plantes. The " Museum " of the Jardin des Plantes is a very remarkable institution, including a zoological and botanical garden, laboratories of chemistry, physics, and physiology, besides the great collections of minerals, fossils, skeletons, and preserved specimens of animals and plants. It is governed by the professors and the director who are in charge of the garden, the laboratories, and the collections, and owes its dignity and its celebrity to the distinguished men of science who for a century and a half have made discoveries and taught there. They are not subject to a board of eminent and wealthy persons, nor is the administration of the antiquities at the Louvre and of the National Library muddled up with that of the great scientific workshop of Natural History.

When the President of the Republic conceived the plan of entertaining the King of Portugal at the Museum of Natural History there were those who supposed that the Minister of Education would, as a great State official, be called upon to arrange the proceedings. Nothing of the sort was done. It was found that the Minister had no authority in regard to the Museum, which, as an independent State institution, organised and carried out the reception through its own officers. The director and professors received President Fallières and the King, escorted by the troops of the Republic. The garden and buildings were ablaze with light and colour, and a large company assembled to take part in the fête. In the great hall of the museum Becquerel, Moissan, and others showed their most recent discoveries as to radium, artificial diamonds, and such matters to the King; others exhibited new birds and fishes, the okapi and newly-discovered fossils, and briefly explained their history and significance. The King conferred decorations on the scientific staff, and gave friendly acknowledgments to all who had thus sought to gratify his special tastes, and prepared for him a really exceptional gala-demonstration of scientific discovery. The official " middle-men," who in other countries contrive to divert the honour and emoluments due to men of science, to their own profit, were on this occasion happily kept at a distance.

36. *The Transmission to Offspring of Acquired Qualities*

The cruel fate of Dom Carlos of Portugal naturally enough produced philosophic and thoughtful articles in some of the journals of the day. An able writer told his readers that the "kingly caste" has characteristics peculiar to itself, "which illustrate the Darwinian law." He does not say what Darwinian law, and I am afraid he would find it difficult to do so. He says that people who for centuries have had their own way (how many kingly families have done so?), who have always lived on good food and never tasted bad wine, and have constantly conversed with interesting people (not usually the chance of princes!) must certainly, if subject to "the laws which govern animal and plant life," produce well-marked characteristics in their offspring—and he goes on to speak of a fine appetite for food (what he describes is really a morbid condition connected with indigestion) as indigenous to Royalty, and declares that the gift of recognising faces and remembering names is "a faculty cultivated by generations of practice."

One must recognise with satisfaction the desire to explain the facts and varieties of human life and character by reference to "the laws which govern animal and plant life." It is by faithfully and truly carrying out the inquiries suggested by that desire that the knowledge which is the sole and absolutely essential condition for the safe conduct of human life and the increased happiness of human communities, can be obtained, and by such inquiries only; and, further, only upon the condition that the investigation is conducted in the true scientific spirit with accuracy and without prejudice. The remarks upon the kingly caste which I have quoted above show with what "legerity and

H

temerity" a clever and respected writer will formulate
phrases and conclusions which are, in face of what
Darwin and his successors have demonstrated, absurdly
erroneous, in fact, topsy-turvy as compared with the
reality.

The main doctrine which Darwin and his followers
have established is that neither castes nor families of
higher or lower living things, including man, acquire
any new characteristics by exposure to special circumstances or by consuming finer or coarser food,
which can or do become innate or fixed in the race.
The individual may be improved or depraved, enlarged
or enfeebled, by the conditions of his individual life,
but he cannot transmit the qualities—the improvement,
the depravity, the enlargement, or the dwindling—
which have been thus attained by him to his offspring.
The race cannot be changed in this way. All the
parents can transmit is the quality which they
themselves have inherited of resisting or of collapsing,
of becoming enfeebled, or of showing strength and
vigour, under certain given conditions. The characteristics of Royalty are not characteristics brought about
by the Royal state, any more than the characteristics
of English race-horses are brought about by the racing
state or by life in a breeder's stable. The characteristics
of Royalty are like those of other living things, the
characteristics of a certain family or blend of families
or strains. Whatever characteristics distinct Royal
families have in common with one another are not due
to the existence of a natural law in virtue of which
the occupations and opportunities of the Royal state
produce " faculties " or " characteristics " in the "blood"
or "stock." Such similarity of characteristics is due
either to the similarity of the demands and conditions
of Court life in all parts of Europe, acting as an
educating force on the individual, or to the inter-
marrying and consequent blending of family character-

istics among a large proportion of the Royal Houses at present existing.

It is very difficult—indeed impossible until much more is written and read on the subjects of breeding and of psychology—to persuade people to abandon the notion that a man who has drunk good wine and conversed with interesting people will, as a direct result, transmit something which he has " taken up " or absorbed from the good wine and the clever people to his offspring, and that a faculty for this or that art or accomplishment cultivated by generation after generation is increased thereby, and transferred as it were into the very vitals of the race—the reproductive germs which each individual has within him. There is no truth whatever in these fancies. They are popular and very natural delusions, which are not only devoid of direct proof by simple observation and experiment, such as that made by all breeders of stock and by medical men, but are also contrary to the great general principles which have been found to explain the varied and most important facts known as to breeding, inheritance, and variation. The same erroneous theory of inheritance now applied to royalty has been put forward in regard to the feeble-minded, the ill-grown, and the incapable at the other end of the social scale.

The only way in which a quality, good or bad, desirable or undesirable, is intensified, made inherent and dominant in a race or strain or family, is by selective breeding—selection due to natural rejection of those individuals not possessing the quality, or to artificial rejection of such individuals by the stock owner and breeder. No human maker of breeds— whether of cattle, horses, birds, or plants—ever yet proceeded by exercising, feeding, educating, or otherwise manipulating his sires and dams ; he simply selects those as parents which by natural variation have the quality, more or less, which he desires, and he destroys or sterilises those which fail to satisfy his requirements. He is perfectly confident that in this way he can ensure the reproduction and exaggeration or dominance of the characteristics which he desires ; he knows that he cannot obtain a " strain " or " breed " by any treatment, any feeding, or education of those which are born without the natural, innate possession of the desired quality, in a more or less marked degree. Once the characteristic turns up as a congenital variation, it can be intensified by coupling its possessor with a mate of like quality ; but both sire and dam have to be rigidly selected with this purpose in view. Such methods are not adopted in human families, even royal ones.

In considering these questions as to characteristic qualities or want of qualities in groups and classes of human communities, we see then that we have in the first instance to distinguish very broadly between the body or structure of the individual, and the " stirps " or germ of the race which he carries within him. The former may be vastly changed for the better or worse as compared with average individuals, without affecting in any way the latter. The germ is carried by the individual member of the race in an almost complete state of isolation or safety from the influences which affect the individual's structure generally (his body as distinct from his germinal or reproductive substance) injuriously or beneficially. The germ varies also, but independently. That is a matter of primary impor- tance. Equally important in the case of man is a peculiarity which affects his manifestation of qualities in a way unknown in any other living thing.

Human society, in more marked and dominating form, in proportion as it is what we call " civilised," has created for itself an inheritance which is not dependent on the variations of strains and the laws of actual breeding. Over and above—very much above—what each man inherits in the form of qualities and characteristics of his special family and stock—is the enormous mass of accumulated experience, knowledge, tradition, custom, and law—which pervades and envelops, as it were, the mere physical generations of this or that pullulating crowd of human individuals. Tradition, at first conveyed by gesture and imitativeness from parents to offspring, then by word of mouth, then by writing, and finally by printed record, sanctioned and enforced by all kinds of persuasion and compulsion —has culminated in an educative discipline which affects every individual in the community in the most powerful way—and constitutes an inheritance of a significance and activity altogether transcending, and independent of that due to the physical transmission of bodily and mental qualities. Public opinion, law, knowledge, belief, custom, and habit exist, and pursue their own course of change, as it were, outside the successive bodily generations of a population. Yet they determine in very large measure the characteristics which each class, and the community as a whole, exhibit. We have to distinguish those results which are due to physical heredity, similar in man and in animals—from results due to this all-powerful education peculiar to man—education, which for civilised man proceeds from almost innumerable sources—from parents, nurses, playfellows, companions, social, professional, and political organisations, as well as from the professed teacher, and from the local peculiarities of the simplest conditions of life. Hence it is that man inherits very little in the way of ready-made instincts, tricks of his nervous mechanism—but, on the contrary, has an enormously long period of individual growth and education, and inherits " educability " to a degree which varies in every family and race.

To estimate correctly, and so to deal with these various factors in human life, we require to know in detail the laws of breeding, heredity, variation, and

selection in animals, and, further, the laws or formulated results of enquiry as to the " educability " of the
human being, the range and the limits of " education,"
the relation of hereditary quality to education, the
causes of mental aberration and defect, of mental
qualities of all kinds, the value and the dangers of all
kinds of educational influences, whether physical, social,
or intellectual.   These are matters in regard to which
there must be in the future more and more of common
knowledge and agreement ; at present they are lightly
touched by politicians and journalists in a way which is
inconsistent with a knowledge of the facts or of their
importance.

When publicists airily declare that the virtues of
kings and the vices of paupers are both due to the
hereditary transmission of characters acquired by the
peculiarities of diet and exercise of the progenitors
of these classes it is time to protest.   To cite the name
of Darwin and " the laws which govern animal and
plant life," in support instead of in condemnation of
such baseless fancies, is, one must suppose, an evidence,
not of a desire to mislead, but of a regrettable indifference to the conclusions of that branch of human knowledge which is of more importance than any other to
the statesman and the philanthropist.

" Selection," whether due to survival in the struggle
for existence or exercised by man as a " breeder " or
" fancier," is the only way in which new characteristics,
good or bad, can be implanted in a race or stock, and
become part of the hereditary quality of that race or
stock.   This applies equally to man and to animals and
plants.   And this selection is no temporary or casual
thing.   It means " the selection for breeding " of those
individuals   which   spontaneously   by   the   innate
variability which all living things show (so that no two
individuals are exactly alike), have exhibited from birth
onwards, more or less clearly, indications of the charac-

teristic which is to be selected. Nothing done to them after birth, and not done to others of their family or race, causes the desired characteristic; it appears unexpectedly, almost unaccountably as an in-born quality. It may be a slight difference only, not easy to take note of; but if it enables those who possess it to get the better of their competitors in the struggle for life, they will survive and mate and so transmit their characteristic to the next generation, whilst those who do not possess it and are beaten in life and fail to obtain food, safety, and mates, will perish and disappear, and their defective strain will perish with them.

37. *Variation and Selection Among Living Things*

Selection is not a thing once done and then dropped —natural selection is continuous and never-ending, except in rare and special circumstances, such as man may bring about by his interference, and then it does not really cease but only changes its demand. The characteristics of a race or species are maintained by natural selection, just as much as they are produced by it. Cessation of a previously active selection (which is sometimes brought about by exceptional conditions) results in a departure of the individuals of the race, no longer subject to that selection, from the standard of form and characteristics previously maintained. To understand this, we must consider for a moment the great property of living things, which is called " variation."

No two animals, or plants even, when born of the same parents, are ever exactly alike. Not only that, but if we look at a great number of individuals of a race or stock, we find that some are very different from the others, in colour, in proportion of parts, in character, and other qualities. As a rule it is difficult to look at such a number, because in Nature only two on the average out of many hundreds, sometimes thousands, born from a single pair of parents, grow up to take their parents' place, and these two are those "selected" by natural survival on account of their close resemblance to the parents. But if we experimentally rear all the off-spring of a plant or animal to full growth—not allowing them to perish by competition for food, or place, or by inability to escape enemies—then we see more clearly how great is the in-born variation, how many and wide are the departures from the favoured standard form which are naturally born and owe their peculiarities to this birth-quality—called innate or congenital variation—and not to anything which happens to them after-wards differing from what happens to their brothers and sisters.

Of course, we are all familiar with this " congenital or innate variation," as shown by brothers and sisters in human families. How and why do innate variations arise ? They arise from chemical and mechanical action upon the " germs " or reproductive cells contained in the body of the parents, and also sometimes from the mating in reproduction of two strains or races which are already different from one another. When an animal or plant is given unaccustomed food or brought up in new surroundings (as, for instance, in captivity) its germs are affected, and they produce variations in the next generation more abundantly. The best analogy for what occurs is that of a " shaking up " or disturbance of the particles of the germ or reproduc-tive material, somewhat as the beads and bits of glass in a kaleidoscope are shaken and change from one well-balanced arrangement to another. And the same analogy applies to the crossing or fertilising of " strain " or " race " by another differing from it. A disturbance is the consequence, and a departure in the form and character of the young from anything arrived at before often takes place. These variations have no necessary fitness or correspondence to the changed conditions which have produced them. They are, so to speak, departures in all and every direction—not very great, but still great enough to be selected by survival if occurring in wild extra-human nature, and obvious enough when produced in cultivated animals and plants to be seen and selected by man, the stock-breeder or fancier.

Indeed the stock-breeder and horticulturist go to work in this way deliberately. Though when they have fattened an animal or fed up a plant they cannot make it transmit its fatness or increased size to its off-spring, yet they can, by special feeding and change of conditions of life—or by cross-breeding—break up the fixed tendency or quality of the germs within the parents so treated. Thus they get offspring produced which show strange and unexpected variations of many kinds—new feathers, new colours, new shapes of leaf, increased size of root, length of limb—all kinds of variations. From the congenital varieties thus produced by "stirring up," "breaking down," or disturbing the germ-matter (germ-plasm) of the parents, the breeder next proceeds to select and mate those which show the character which suits his fancy, whilst he destroys or rejects the others. Thus he establishes, and by repeated selection in every generation maintains, and if he desires increases, the characteristics which he values.

Birth-variation is then an inherent property of living things (including man) as much as heredity, which is the name for the property expressed in the resemblance of offspring to parent. And birth-variation, or congenital variation—that is to say, the being born with a power to grow into something different (not greatly, but still obviously, different) from their parents or ancestry, and from their brethren and cousins, though not subjected after birth to any treatment or conditions differing from those common to all of them—is a quality of living things which must be distinguished altogether from the power of the individual itself, though not born with qualities differing from those of its brothers and sisters, to vary or change in some respects as compared with other individuals when it is specially fed or exposed to special treatment. The first is change, or variation, of the " stirps," or germ plasm ; the second is change, or variation, of the transient body of the individual. The first is indefinite and may be of almost any kind or form ; once it has appeared, it is a permanent possession of the race descended from its owner. The second is definite and a direct reaction to the environment. Such an individually induced or stimulated change is often called an " acquired character." It does not affect the stirps, the inner reproductive germs, and cannot be handed on by inheritance to a new generation.

What happens, then, when there is a cessation of selection ? All sorts of birth-variations appear and grow up. The fine adjustment of form—maintained by natural selection carried on unceasingly—no longer obtains. The characteristics of the race become less emphasised. All sorts of birth-variations have an equal chance, and the tendency must be for those characteristics which have most recently been established and maintained by severe selection to dwindle and then to disappear altogether. The majority of birth-variations will—when selection is prevented—always tend to present a lessened, rather than an increased, development of any one characteristic—the excelling minority will no longer be selected, but all will have an equal chance in mating and reproducing. Hence, bit by bit, all salient features, all the characteristics of the race previously maintained by selection, will, as a result of survival of all variations and general crossing and interbreeding—dwindle and disappear. It is to this process that the term "degeneration" has been applied by biologists. How far it may go, and what are its limits and various outcomes, I cannot now discuss. It is sometimes spoken of as "retrogression" —which implies wrongly a return to a previous state. From some points of view it might be called "simplification."

The point to which I have been making is this—that civilised mankind appears to be very nearly in regard to most points of structure and quality in a condition of "cessation of selection." It is the better-provided and well-fed, well-clothed, protected classes of the community, in which this cessation of selection is most complete. Racial degeneration is, therefore, to be looked for in those classes quite as much as in the half-starved, ill-clad, struggling poor, if, indeed, it should not be expected to be more strongly marked in them. There are facts which tend to show that such anticipations are well-founded.

This is a matter requiring further discussion. It is probable, I may say in anticipation, that whilst natural selection in the struggle for existence is only obscurely operative (except as to alcoholism and some diseases) in civilised man, yet what Mr. Darwin called sexual selection—the influence of preference in mating—has an important scope, and it may be that hereafter it will be of enormous importance in maintaining the quality of the race.

Meanwhile, it seems that the unregulated increase of the population, the indiscriminate, unquestioning protection of infant life and of adult life also—without selection or limitation—must lead to results which can only be described as general degeneration. How far such a conclusion is justified, and what are possible modifying or counteracting influences at work which may affect the future of mankind, are questions of surpassing interest. In any case, it is interesting to note that the cessation of selection is more complete, and the consequent degeneration of the race would, therefore, seem to be more probable in the higher pro-pertied classes than in the bare-footed toilers, whose ranks are thinned by starvation and early death. One may well ask, " Is this really so ? "

38. *The Movement, Growth, and Dwindling of Glaciers*

Last summer we were watching the gradual change of the brilliant sunlight on the snows of Mont Blanc as the shadows crept up the pine-covered sides of the valley of Chamonix. We noted how the highest peak —the true summit of Mont Blanc—remained almost white and brilliant when the somewhat lower and nearer Dome de Gouter (so often, when clouds are about, mistaken for the true summit by tourists) had assumed a marvellous shade of saffron-rose colour. The crevasses of the glaciers were marked by an unearthly pale-green tint and delicate purple hues of weird beauty were spreading over the evanescent forms of the great snow-field, when one of the hotel guests—a citizen of Geneva—said, " Ah, yes ! Look at them whilst you may, and wonder at them, those glaciers of the Alps. They are but the remnants, the roots, as it were, of the vast glacier which once filled the whole of this vale of Chamonix and spread down into the valley of the Rhone, and ploughed out with the slow movement of its huge mass the deep rock basin of the Lake Leman. Every year they dwindle, as they have dwindled for ages past, and soon—perhaps not more than another 100 years hence—they will have disappeared utterly from human sight and knowledge." I continued to gaze at the scene, and as the night fell and the distant details were lost to view I felt as though a venerable, but decrepit, friend had passed from my sight, never to return. I was rejoiced to see the glaciers still there when the morning sun showed forth their strange opaque white and faintly green masses on the mountain sides—stupendous outpourings, as it were, of whipped cream tinted with pistachio-nut.

But was it true, that lament of the Genevese savant? Undoubtedly the glaciers in many parts of the Alps have been shrinking for the last thirty years. It is longer than that since I first saw the glaciers of the Chamonix valley, and there is no doubt that they have shrunk up since then, leaving acres of boulders and bare polished rock where was the ice I formerly climbed. The glacier of Argentière, near the upper end of the valley, is a mile or more shorter than it was; the ice caves which we used to visit at the foot of the Mer de Glace have melted away, and the end of the glacier is now high up above a precipitous surface of polished rock far from the site of the little pavilion, with its gay flag and amiable guardian, who used to exhibit the marvellous ice cavern.

I find on looking into the matter that it is true that there has, during the latter half of the past century, been a great dwindling of the lower end or " snout," a drawing back, as it were, not only of Swiss glaciers, but of glaciers in other parts of the world—as, for instance, in Alaska and in the Himalayas. But I cannot avoid a feeling of satisfaction in recording the opinion of geological authorities that, contrary to the assertion of the Swiss pessimist, there is not any ground for believing that the present noticeable shrinking is due to a continuous process by which the enormous glaciers of remote ages have been incessantly reduced until now they are but rootlets or stumps of the former masses, destined to evaporate completely under the continued remorseless operation of increasing temperature. On the contrary, it appears that, though there are not accurate records and measurements as to past centuries as there will be as to present and future years, yet there is abundant evidence that Alpine glaciers have grown longer in some centuries and retreated in others. The period of alternate extension and retraction has not been ascertained with accuracy, but by some geologists it is supposed to be about fifty years. The retraction or shrinking is not due to a continuous increase of the temperature of the earth's atmosphere—or of this hemisphere—but to contending causes which operate alternately towards increase and towards decrease when one or two hundred years are considered. Such are the greater or less rainfall and snowfall over a very large area, and the formation and persistence of clouds, concerned with which are probably those varying quantities—the spots on the sun.

The simple proof that glaciers have extended and again retreated within historic times is furnished by the fact that in some parts of the Alpine range the retreat of a glacier has uncovered ancient miners' excavations, which must have been worked when the glacier did not reach the spot excavated. Subsequently the glacier advanced, and now after some hundreds of years it has again retreated and exposed the ice-covered borings and workings. The tradition of a glacier-enclosed village in the Zermatt mountains, shut off from the world by the advance of glaciers, lost and mysterious, is evidence that such advance has been observed by the native population.

The natives who live near glaciers know that they advance and retreat, but the fact that the whole glacier is really a slowly flowing viscous mass—a sort of frozen but not immobile river—was only established by scientific observation in the last century. The frozen river is fed by the snow which falls on the higher mountain ridges, and is squeezed into the form of ice instead of snow powder by its own weight as it slips down the inclines, warmed by the unclouded sunshine. The big glaciers move much more rapidly (or perhaps one should say less slowly) in the middle than at the sides. The measurements which have been made differ in different glaciers and in different parts of the same glacier, and show smaller movement in winter than in summer. The advance of the sides is retarded, as in the case of an ordinary river of flowing water, by friction against the rocks, which enclose the glacier as its banks enclose a river. A good average case shows a flow downwards in summer of half a foot a day at the sides and a foot and a half in the middle. The distance below the snow-line to which the flowing glacier descends down a mountain gorge—before it melts away and becomes a river of liquid water—depends, as does the rate at which it moves, in the first place, on the temperature of the region and on the sharpness of the slope. A glacier will flow downwards (as will a lump of pitch) along a scarcely perceptible incline, but more slowly than down a steeper incline, and it will, consequently, get further down into the warm valley without altogether melting away when the slope is steep.

But apart from these considerations, the bigger and thicker (or deeper) the glacier, that is to say, the more snow which each year falls at its starting-place and goes to making it, the further down will it flow before melting away ; and it is the heavy snowfall of many years ago or of a series of years long past which has to-day reached in the form of ice the lower end of the glacier. So, though the lower end of the glacier may melt more quickly if the valley has become hotter, yet the heavy snowfalls of fifty years ago may only now have reached the valley, and may quite counterbalance the melting action of the warmer summers. Or reverse conditions, namely, less snow and lower or unchanged temperature in the valley, may prevail.

The Government of India has lately established a definite survey and record of the movement of several Himalayan glaciers and of the variation in the distance to which their "snouts" descend into the valleys. Twelve glaciers were examined last year, and will be properly watched in future. The Yengutsa glacier has gained about two miles in length since Sir Martin Conway visited it in 1892; the great Hispar glacier has slightly retreated. The Hassanabad glacier three years ago increased its length by a rapid progress of the free "snout" of as much as six miles in three months, and is now no longer increasing or advancing! Many years ago it had reached its present position, and then retreated. The rock masses carried on the ice and left in great heaps at the point where the glacier melted away are known as terminal "moraines," and often serve to show the position to which the snout of a glacier once extended—far below its present limit. A curious fact as to the increase and shrinkage of glaciers is that of two neighbouring glaciers, as in the case of the glacier Blanc and the glacier Noir in Dauphiné (France), one may be advancing whilst the other is in retreat. Further study and knowledge of the causes of these variations will throw important light on questions of general meteorology.

Although there is no evidence to lead us to suppose that existing glaciers are now actually in a condition of general retreat, leading to their ultimate disappearance, yet it is one of the most certain and interesting results of geological study that some hundred and fifty thousand years ago the northern hemisphere was far colder than it is now, owing partly to the same change in the inclination of the earth's axis to which I alluded on a former page (p. 81) as affecting the orientation of ancient astronomical temples—a change which diminished, when at its extreme, the effective amount of heat received from the sun in these regions of the earth. The peculiar scratching, polishing, and erosion of rocks, the existence of moraines, and other evidence, prove that enormous glaciers covered the north of Europe, that England and Scotland were in large part covered by a great ice-sheet or glacier, and that the great valleys of Switzerland such as the Rhone Valley and the basin of the Lake of Geneva, were filled by enormous glaciers, which helped to mould and deepen the valleys. The present glaciers are truly the remnants or rootlets of those enormous masses of the glacial epoch. On such of the land surface as was not then covered by ice, existed the hairy elephant or Siberian mammoth, the woolly rhinoceros, wild cattle, lions, bears, hyenas, and other animals now extinct in this part of the world. Man had made his appearance, hunted these animals, and lived in caves. His weapons and carvings and their bones tell us the story in no uncertain terms.

The biggest Swiss glaciers of to-day, compared to the great glacier of the Rhone Valley, of which they are but the highest tributaries, still surviving unmelted among the mountain-tops, are in size as a mountain freshet is to the great stream of Loch Lomond, or as the Serpentine in Hyde Park to the neighbouring Thames. Vast as was the great glacier of the Rhone Valley, and immense as has been the work done by water and ice in carving the great highway in the mountain-mass of Switzerland, it has all been effected since the date of the formation on the sea-bottom and the subsequent elevation of the strata which we call " the chalk "—a deposit which comes not very far down in the series of strata of the earth's crust. Only 3,000ft. of deposit exist above it, whilst below it are more than 60,000ft. of water-deposited. or " sedimentary " rocks. The huge Alps have risen since the date of the " chalk," for we find strata containing marine shells of the Tertiary period at a height of 10,000ft. in those mountains. Where those shells now are was the bottom of the sea at a comparatively recent date, probably not more than fifty million years ago! And not only have the Alps been raised since then from the sea level to 15,000ft. (the height of Mont Blanc), but the huge mountain valleys and the great chasm of the Rhone Valley many miles wide, with its floor thousands of feet below the mountain ridges, have been scoured out. Deeper and wider it has gradually become as it has taken shape, whilst the mountain sides have been removed first by water and later by ice—by the great glacier consisting of solid ice, miles wide and a thousand and more feet in thickness. The water no longer fills the valley in solid form, but once again rushes along as an irresistible torrent, tearing and wearing the rock without rest or mercy, carrying it off by thousands of tons day by day, year by year, to the plains of Provence and the deep floor of the Mediterranean Sea.

The blue colour of the glacier ice—like that of pure water—is now known to be due to no impurity or admixture of other substances. It does not, as was supposed by Tyndall, owe its blueness to a dust of finest colourless particles as do blue smoke, the blue sky, and as do the blue eyes which have attracted the observation of naturalists (and others) in Ireland and the North of Europe. Water, whether liquid or solid, is blue, just as " blue copperas " is, or as " Prussian blue " is; but light must pass through some ten or twenty feet thickness of it to make the colour evident to our eyes. The green tint is due to an admixture of yellow, the exact cause of which is not quite easy to discover. Probably it is due to minute quantities of earthy matter mixed with the surface snow.

The pressing of the high-lying snow, so as to form solid ice or "glacier," is concerned with the same property of snow as that which enables us to make snow

" bind" into a snowball. You cannot make snowballs during very hard frost—the snow must be in air of a thawing temperature at the moment it is squeezed by the hand. The hand itself will not be warm enough to produce that temperature when the thermometer is below freezing-point. The snow commences to melt in the hand when one squeezes it, and then when the squeezing is stopped the water formed quickly freezes again and cements the snow particles together to form ice, enclosing innumerable minute bubbles. The heat of the sun and the pressure of the weight of the snow itself take the place in the mountains of the warmth and pressure of the human hand. The minute air bubbles make the newest glacier-ice white and opaque, especially when seen in a great mass; but gradually they get squeezed together, and the glacier ice becomes first "fibrous" in appearance, and then, after long years of pressure by its own weight, fairly clear. Ice in great masses has the properties of a viscous body, like pitch or soft sealing-wax, owing to the fact that wherever the solid mass breaks its particles melt a very little and then freeze again. Under increased pressure ice melts at a lower temperature than when it is not subjected to pressure. When the pressure is removed the water freezes again. Thus crushed ice or snow can be put into a "squeeze-mould" and pressed, so as to form a solid mass of ice of any shape you may choose. Four or five slabs of ice, placed one over the other, very soon become, owing to this property, one continuous solid mass. White glacier ice is so full of air bubbles as to be comparable in structure to sponge, or, more closely, to cork. A cube of such ice exposes, owing to its rough air-hole pitted surface, a much larger surface of contact to the atmosphere than does a cube of perfectly smooth clear ice. Consequently in a warm room or chamber the white ice melts much more quickly than does the clear, and hence you should choose clear ice rather than white ice if you wish for a block which will last.

Before leaving the glaciers, let me briefly relate an incident arising from their slow but regular downward flow to the region where they melt away and deposit, as a terminal moraine, the burden of rocks they have received years before in regions far above. A young man of five-and-twenty, on his honeymoon, visited the Alps, and ventured alone on to a glacier. He fell into a deep "crevasse," or ice-fissure, and his body was not recovered. The exact spot where he fell into the ice-chasm was recognised, and the mountain-folk, who knew their glacier and its rate of movement well, told the broken-hearted young widow that it would take thirty years before that region of the glacier would have moved so far downwards as to reach the lowest limit, and in due course melt away. She haunted the glacier in which her young husband was entombed year after year, and at last, when she was now grey-headed and withered by time, that special tract of ice had descended so far, and was so near the thawing, thinned-out margin of the glacier that they were able to break into it with axe and pole. Then she, an old woman, had a wonderful experience. They led her to the glacier's edge. Her young husband, pre-served these thirty years in the ice, which had melted around him and re-frozen, lay there unchanged. His features were not marred by the lapse of years, nor was his clothing rent or injured. He seemed as one asleep, resting after a long day's climb, and she, poor soul, had, during a blissful interval, the conviction that all those weary years of waiting were but a long, bad dream, that she, too, still was young, and was waking, as she had loved to do long years ago, in time to see him lift his lids and smile.

39. Votes for Women

Now that so many people placidly accept the notion that women are to have votes in the election of members of Parliament, one is tempted to ask whether science has any facts to put forward which should be considered before so great a change in our national organisation is made. There are various interesting facts as to the relations of males and females in the animal world and as to the relative strength and activity of the sexes—which are sometimes cited as arguments in the matter. Speaking generally, it is clear enough that among animals the female is endowed with qualities which bear exclusively upon her function as the guardian of the eggs or germs of a new generation. She nourishes those germs at the expense of her own substance before birth, feeds them, tends them and protects them—after birth. The male in many cases contributes to the feeding and protection of the young, but is as often as not quite unconcerned with such matters. In the higher animals the male is far more powerful than the female, and fights with other males both for the possession of a mate or a harem, and for the undisturbed occupation of feeding grounds for himself and family.

Among lower animals there are curious cases of the greater strength and size of the female. Thus, among spiders, the female is nearly twice as bulky as the male. She makes, in many cases, a nest ready for her young, and is visited there by the wandering irresponsible male, who, in spite of great danger to himself, is irresistibly attracted to seek a brief caress from the terrible spideress. She is terrible, not only on account of her bulk, but because she makes a rule of killing, and sucking the blood of, her infatuated admirer unless he is sufficiently alert and agile to escape from her side more quickly than he came to it. The courtship of spiders is a very interesting bit of natural history. The males execute a sort of dance, and are strangely excited by the vibrating note of a tuning fork. Two American naturalists, Mr. and Mrs. Peckham, and also Dr. McCook, have studied this subject in great detail.

A strange-looking, dark green worm, as big as a walnut, with a ribbon-like trunk six or eight inches in length attached to its mouth, lives in holes in the rocks in the Mediterranean. A similar worm has been found off the Norwegian coast. Fanciful names are given by zoologists to these two worms—the first is called Bonellia, the second Hamingia. It does no harm to cite their names, and I do so with an apology to those who do not like names. These goodly sized worms are females, only females. For years the corresponding male was unknown. At last a minute creature one-eighth of an inch in length, like a tiny fragment of green thread, was found crawling about on and into these big green Bonellias. Its structure when it was examined with the microscope proved it to be the adult male of the worm on which it was crawling. It was so insignificant and minute as to escape all observation except that of a trained naturalist searching for it with a magnifying glass. Some seven or eight of these diminutive males are found on one female, infesting her as fleas infest a mouse, and of about the same relative size. The microscopic husband of the Norwegian Hamingia it was my good fortune to discover many years ago, when I was dredging marine animals in the deep waters of the Stavanger Fjord.

So there is nothing in the eternal fitness of things proclaiming the male as the necessary superior of the female throughout Nature. The fact is that the question of equality and of general superiority and inferiority has no place in regard to male and female from a naturalist's point of view. It is true that women are so very much less endowed with muscular strength than men that practically every woman is inferior to every man in this respect.   It is also true that woman's brain is smaller than man's, and that apart from mere size, the intellectual activity and capacity of women, by whatever test you examine it, is less than that of man.   When exceptional cases on both sides are excluded, the definite intellectual inferiority of the average woman, as compared with the average man, is established as a fact. The observations of those concerned in the education of young men and young women side by side confirm this, and it is further demonstrated by a consideration of the intellectual performances of average men and average women.   That, at any rate, is my own experience as a University teacher.   But women, on the other hand, fill a place in human life as mothers, and administrators of detail, and as companions, in which man, by the nature of things, cannot compete with them at all.

At the house of the late Sir James Knowles, some twenty-five years ago, when discussing the relative value of the physical and intellectual capacities of the men as compared with the women of the English working class, Mr. Gladstone (at that time the head of the Government) said to me, "I am of opinion that the relative value of a man and a woman is in all classes of society about the same as it was in my grandfather's time in Jamaica when they purchased slaves.   They gave £120 for a man and £80 for a woman, and that is a fair measure of their relative value all the world over."   It is necessary to remember that Mr. Gladstone was not estimating the ultimate value of woman in human life when he said this.   He would, I think, have considered, as I do, that it is absurd to attempt to estimate that or to raise a discussion as to general superiority and inferiority in reference to the male and the female of the human species.   They are creatures as necessary one as the other, differing from one another profoundly and excelling one another in diverse qualities and capacities. Without this complementary division of fitness and quality our life would be a monotone robbed of the infinite variety which characterises humanity. What Mr. Gladstone estimated as being less by one-third in women than in men is power—work-value—whether physical or intellectual. I think Mr. Gladstone's estimate must be admitted as true.

But I do not for a moment say that when this inferior intellectual and physical capacity of woman is admitted the question is settled as to whether women should vote for the election of representatives to carry on the affairs of the country. The affairs of the country! They are, in the first place, the protection of person and property by the law, which must be upheld by force if necessary; then defence against foreign aggression, also a matter of force; and, further, the education and training not only of children but of the ripe youth of the country—a matter of intellect—which also has a weighty influence in the making of wise laws. Then there is the devising of weapons and means of defence by land and by sea, as well as the discovery and application of knowledge in regard to disease, both of mind and body, for the benefit of the community. And there will soon be a good deal more!

It does not necessarily follow, because women cannot themselves do some of these things at all, and for the others are less able than men, that they should not give a vote in electing the men who are to attend to them. The only question is, Would it make life better for both women and men were they allowed to do so?

The argument that the paying of taxes on men's property qualifies men to give a vote, and therefore the paying of taxes on women's property should, *ipso facto*, entitle women to give a vote, is fallacious, because the paying of taxes is not the reason or determining cause of men having a vote, but only a subsidiary test or qualification which might be abolished or modified. The property of minors pays the tax, but it is not proposed on that account that children should vote. The property qualifications in use at present are merely a method for excluding certain men, and we might have an intellectual qualification or a muscular qualification for the same purpose. Indeed, we do at present exclude male imbeciles and those who are immature. The reason for extending the Parliamentary vote to a larger and larger body of the male population has been to secure the assent of the strength and manhood of the country to the laws and public acts of the Government, and to ensure its willing participation in that maintenance of the central Government's decisions by physical force, which is the ultimate and by no means very remote method by which they are maintained. It does not seem to be likely to be an improvement on our present system that women, who must always be regarded as specially privileged because of their physical weakness, should nevertheless be allowed to influence by the mere number of their votes the decision of questions in which the employment of the physical strength of men acting as defenders of our territory, guardians of the peace, or ministers of the law, is the essential condition of an effective result following on such decision.

To a naturalist human population does not appear as a number of units of which a few more are female than male—but rather as a series of families, consisting of men, women, and children, bound together by a variety of reciprocal services, dependent one on another, ordered and disciplined to a distribution of functions and duties by the tradition and experience of ages. The notion that the paterfamilias is the rightful chief of his wife and children, and that through him they are represented, and should be content to be represented, in the local and greater State Government—is one of long standing in civilised Europe. The powers of the paterfamilias have been gradually limited in the course of the development of social life since the young men and the old bachelors, too, have been given a share of power in the State : but the recent proposal to break the fabric of his household by giving the Parliamentary franchise to women is so sudden and strange a notion that he seems not to have realised what it means.

The apathy which many men exhibit in regard to this proposal is as remarkable as the amiable courtesy with which others assent to it rather than " disoblige a lady." Looking at the proposal not as a question of justice, which really has nothing to do with it, but in reference to the inquiry as to whether it is likely, if carried, to increase the happiness and prosperity of the community, I must say that, so far as the natural history of man gives indications, it seems to me that if women acquired the Parliamentary franchise and made active use of it, they would be led into a new attitude of independence and separation from the men and from the family group to which they are by birth or alliance attached. I fear that the great business of making the nest beautiful, producing and tending the young, nursing the sick, helping the aged, consoling the afflicted, re-warding the brave, of dancing and singing and creating gaiety within the charmed circle where political contests and affairs of State are of no account, would be ne-gleeted and without honour. In the end these amenities of life would probably fall into the hands of commercial companies and be sent out at so much a head—im-ported from Germany. Woman would not be the gainer, for she can only gain by continuing to astonish man by all she does for his enchantment and delight, to serve him and to crown his life—she will only suffer by becoming "independent." The movement which is supposed to lead to a higher development of woman-hood, and consists in women mobbing people on their doorsteps, waving flags and shouting at other people's meetings, and struggling in the arms of policemen, seems to be inconsistent with a development in the direction which has hitherto been popular and successful in the progress of man from savagery to decency. It is difficult to suppose that men will really be so blind to the facts of the real importance and true value of women as to allow this movement to succeed whilst they look on with vague incredulity as to its being anything more than a huge joke. .

There is, too, finally, one serious warning to be derived from the ascertained facts of human physiology and psychology. The immutable task, the sacred destiny, of women is to become the mothers of new generations. Nothing which is likely to interfere with or lessen the respect and veneration due to women in view of this tremendous natural determination of their instincts and aspirations should be lightly sanctioned by men so long as they have the power of deciding the matter. There is good and sufficient ground for fearing that the new status of women which would be established by their entry on an equal footing with man into the arena of political struggle and public life, would injuriously affect in a majority or large minority of cases that mode of life and economy of strength which is necessary for those who must give so much to the great and exacting demands of maternity. The gratification of the whim of a few earnest but injudicious women would be an altogether insufficient justification for the injury of the " physique " of women in general by the strain of public competition with men, and for the widespread development in women of an increased habit of self-assertion and self-sufficiency— habits which must make them unwilling to accept their natural duties as wives and mothers—and must make men equally unwilling to promote them to these honours and privileges.

40. *Tobacco and the History of Smoking*

A proposal is before Parliament to prevent little boys from "smoking" in public places. Little girls are, as the bill at present stands, not to be interfered with. Perhaps this is because they are not to have votes when they grow up, and so they may do as they like.

Apart from the question as to whether the smoking of tobacco is injurious to the health or not, there are many curious questions which arise from time to time as to the history and use of tobacco. I have no doubt that for children the use of tobacco is injurious, and I am inclined to think that it is only free from objection in the case of strong, healthy men, and that even they should avoid any excess, and should only smoke after meals, and never late at night. The strongest man, who can tolerate a cigar or a pipe after breakfast, lunch, and dinner, may easily get into a condition of " nerves " when even one cigarette acts as a poison and causes a slowing of the heart's action.

A curious mistake, almost universally made, is that of supposing that the oily juice which forms in a pipe or at the end of a cigar is " nicotine," the chief nerve-poison of tobacco. As a matter of fact, this juice, though it contains injurious substances, contains little or no " nicotine." Nicotine is a colourless volatile liquid, which is vapourised and carried along with the smoke ; it is not deposited in the pipe or cigar-end except in very small quantity. It is the chief agent by which tobacco acts on the nervous system, and through that on the heart—the agent whose effects are sought and enjoyed by the lover of tobacco. A single drop of pure nicotine will kill a dog. Nicotine has no aroma, and has nothing to do with the flavour of tobacco, which is due to very minute quantities of special volatile bodies similar to those which give a scent to hay.

Most people are acquainted with the three ways of "taking tobacco"—that of taking its smoke into the mouth, and more or less into the lungs, that of chewing the prepared leaf, and that of snuffing up the powdered leaf into the nose, whence it ultimately passes to the stomach. A fourth modification of the snuffing and chewing methods exists in what is called the "snuff stick." According to the novelist, Mrs. Hodgson Burnett, the country women in Kentucky use a short stick, like a brush, which they dip into a paperfull of snuff; they then rub the powder on to the gums. Snuff-taking has almost disappeared in "polite society" in this country within the past twenty years, but snuffing and chewing are still largely practised by those whose occupation renders it impossible or dangerous for them to carry a lighted pipe or cigar —such as sailors and fishermen and workers in many kinds of factories and engine-rooms.

One of the most curious questions in regard to the history of tobacco is that as to whether its use originated independently in Asia or was introduced there by Europeans. It is largely cultivated and used for smoking throughout the East from Turkey to China —including Persia and India on the way—and special varieties of tobacco, the Turkish, the Persian, and the Manilla are well known, and only produced in the East, whilst special forms of pipe, such as the "hukah" or "hooka," the "bubble-bubble," and the small Chinese pipe are distinctively Oriental. Not only that, but the islanders of the Far East are inveterate smokers of tobacco, and some of them have peculiar methods of obtaining the smoke, as, for instance, certain North Australians who employ "a smoke-box" made of a joint of bamboo. Smoke is blown into this receptacle by a faithful spouse, who closes its opening with her hand and presents the boxful of smoke to her husband. He inhales the smoke and hands the bamboo joint back to his wife for refilling. The Asiatic peoples are great lovers of tobacco, and it is certain that in Java they had tobacco as early as 1601, and in India in 1605. The hookah (a pipe, with water-jar attached, through which the smoke is drawn in bubbles) was seen and described by a European traveller in 1614. Should we not, therefore, suppose that in Asia they had tobacco and practised smoking before it was introduced from America into the West of Europe? It seems unlikely that Western nations should have given this luxury to the East when practically everything else of the kind has come from the East to Europe—the grape and wine made from it, the orange, lemon, peach, fig, spices of all kinds, pepper and incense. Yet it is certain that the Orientals got the habit of smoking tobacco from us, and not we from them.

Incredible as it seems, the investigations of the Swiss botanist, De Candolle (see his delightful History of Cultivated Plants—a wonderful volume, published for 5s., in the International Scientific Series) and of Colonel Prain, formerly in India, now Director of Kew, have rendered it quite certain that the Orientals owe tobacco and the habit of smoking entirely to the Europeans, who brought it from America, as early as 1558. In the year 1560 Jean Nicot, the French Ambassador, saw the plant in Portugal, and sent seeds to France to Catherine de' Medici. It was named Nicotiana in his honour. But the introduction into Europe of the practice of smoking is chiefly due to the English. In 1586 Ralph Lane, the first Governor of Virginia, and Sir Francis Drake brought over the pipes of the North American Indians and the tobacco prepared by them. The English enthusiasm for tobacco smoking, "drinking a pipe of tobacco," as it was at first called, was extraordinary both for its sudden devolopment, its somewhat excessive character, and the violent antagonism which it aroused, and, as we learn from Mr Frederic Harrison, still arouses. It was at once called "divine tobacco" by the poet Spenser, and "our holy herb nicotian" by William Lilly, and not long afterwards denounced as a devilish poison by King James. The reason why the English had most to do with the introduction of smoking is that the inhabitants of South America did not smoke pipes, but chewed the tobacco, or took it as snuff, and less frequently smoked it as a cigar. From the Isthmus of Panama as far as Canada and California, on the other hand, the custom of smoking pipes was universal, and wonderful carved pipes of great variety were found in use by the natives of these regions, and also dug up in very ancient burial grounds. Hence the English colonists of Virginia were the first to introduce pipe-smoking to Europe.

The Portuguese had discovered the coasts of Brazil as early as 1500, and it is they who carried tobacco to their possessions and trading ports in the Far East—to India, Java, China, and Japan, so that in less than a hundred years it was well established in those countries. Probably it went about the same time from Spain and England to Turkey, and from there to Persia, and rapidly developed not only special new forms of pipe (the hookah) for its consumption, but also within a few years special varieties of the plant itself. These were raised by cultivation, and have formerly been erroneously regarded as native Asiatic species of tobacco plant.

The definite proof of the fact that tobacco was in this way introduced from Western Europe to the Oriental nations is, first, that Asiatics have no word for it excepting a corruption of the original American name tabaco, tobacco, or tambuco: it is certain that it is not mentioned in Chinese writings nor represented in their pottery before the year 1680. In the next place, it appears that careful examination of old herbariums and of the records of early travellers who knew plants well and recorded all they saw, proves that no species of tobacco is a native of Asia. There are fifty species of tobacco, but all are American excepting the Nicotiana suaveolens, which is a native of the Australian continent, and the Nicotiana fragrans, which is a native of the Isle of Pines, near New Caledonia.

Forty-eight different species of tobacco (that is to say, of the genus Nicotiana) are found in America. Of these Nicotiana tabacum is the only one which has been extensively cultivated. It has been found wild in the State of Ecuador, but was cultivated by the natives both of North and South America before the advent of Europeans. It seems probable that all the tobaccos grown in the Old World for smoking or snuffing are only cultivated varieties—often with very special qualities—of the N. tabacum, with the exception of the Shiraz tobacco plant, which, though called N. persica, is of Brazilian origin, and the N. rustica, of Linnæus, a native of Mexico, which has a yellow flower, and yields a coarse kind of tobacco. This has been cultivated in South America and also in Asia Minor. But tobaccos so different as the Havannah, the Maryland and Virginian, the incomparable Latakia, the Manilla, and the Roumelian or Turkish—all come from culture-varieties of the one great species, Nicotiana tabacum.

The treatment of tobacco-leaf to prepare it for use in smoking, snuffing, and chewing requires great skill and care, and is directed by the tradition and experience of centuries. As is the case with " hay," the dried tobacco-leaf undergoes a kind of fermentation, and, in fact, more than one such change. The cause of the fermentation is a micro-organism which multiplies in the dead leaf and causes chemical changes, just as the yeast organism grows in " wort " and changes it to " beer." It is said that the flavour and aroma of special tobaccos is due to special kinds of ferment, and that by introducing the Havannah ferment or micro-organism to tobacco-leaves grown away from Cuba, you can give them much of the character of Havannah tobacco! A very valuable kind of tobacco is the Roumelian, from which the best Turkish cigarettes are made. It has a very delicate flavour, and very small quantities of an aromatic kind prepared from a distinct variety of tobacco plant grown near Ephesus and on the Black Sea (probably a cultivated variety of Nicotiana rustica) are judiciously blended with it. This blending, and the use of the very finest qualities of tobacco-leaf, are essential points in the production of the best Turkish cigarettes. The so-called " Egyptian " cigarettes are made from less valuable Turkish tobacco, with the addition of an excess of the aromatic kind. It is a mistake to suppose that opium or other matters are used to adulterate tobacco. The only proceeding of the kind which occurs is the mixing of inferior, cheap, and coarse-flavoured tobaccos with better kinds. Water and also starch are used fraudulently to increase the weight of leaf-tobacco. But skilful " blending " is a legitimate and most important feature in the manufacture of cigars, cigarettes, and smoking mixtures.

The first "smoking" of tobacco seen by Europeans was that of the Caribs or Indians of San Domingo. They used a very curious sort of tubular pipe, shaped like the letter Y. The diverging arms were placed one up each nostril, and the end of the stem held in the smoke of burning tobacco-leaves, which was thus "sniffed up" into the - nose. The North American Indians, on the other hand, had pipes very similar to those still in use. The natives of South America

K

smoked the rolled leaf (cigars), chewed it, and took it as snuff.

It has been suggested that in Asia smoking of some kind of dried herbs may have been a habit before tobacco was introduced—since even Herodotus states that the Scythians were accustomed to inhale the smoke of burning weeds, and showed their enjoyment of it by howling like dogs ! But investigation does not support the view that anything corresponding to individual or personal " smoking " existed. " Bang " or " hashish " (the Indian hemp) was not " smoked," but swallowed as a kind of paste before the introduction of tobacco-smoking in the East—as we may gather from the stories of the " Arabian Nights "—although the practice of smoking hemp (which is the chief constituent of " bang ") and also of smoking the narcotic herb " henbane," has now been established. Opium was, and is, eaten in India, not " smoked." The " smoking " of opium is a Chinese invention of the eighteenth century.

The Oriental hookah suggests a history anterior to the use of tobacco, but nothing is known of it. The word signifies a cocoanut-shell, and is applied to the jar (sometimes actually a cocoanut) containing perfumed water, through which smoke from a pipe, fixed so as to dip into the water, is drawn by a long tube with mouthpiece. It seems possible that this apparatus was in use for inhaling perfume by means of bubbles of air drawn through rose-water or such liquids, before tobacco-smoking was introduced, and that the tobacco-pipe and the perfume-jar were then combined. But travellers before the year 1600 do not mention the existence of the hookah in Persia or in India, though as soon as tobacco came into use this apparatus is described by Floris, in 1614, and by Olearius, in 1633, and by all subsequent travellers.

The conclusion to which careful inquiry has led is

that though various Asiatic races have appreciated the smoke of various herbs and enjoyed inhaling it from time immemorial, yet there was no definite " smoking " in earlier times. No pipes or rolled-up packets of dried leaves—to be placed in the mouth and sucked whilst slowly burning—were in use before the introduction of tobacco by Europeans, who brought the tobacco-plant from America and the mode of enjoying its smoke, and passed on its seeds to the people of Turkey, Persia, India, China, and Japan.

41. *Cruelty, Pain and Knowledge*

It is difficult to write or to read or even to think about " cruelty " and preserve one's sober judgment and reason. Most people are upset by emotion when torture and the details of the infliction of pain are discussed All the more must we remember that emotion is a powerful driving force, but a bad guide. Only true knowledge and sound reasoning can guide us aright.

An awful fact about the emotional state produced by witnessing or hearing about the agonies of human beings or of sentient animals is that to some people (actually very few and diminishing in number among civilised races) it is distinctly a source of pleasure, though to most of us it is intolerably painful. This fact forms one of the most difficult problems of psychology. It seems that just as there are people who enjoy seeing dangerous acrobatic performances or climbing themselves among ice and rocks at the risk of their lives, or reading of hairbreadth escapes, of bloody murders, of ghosts, and other horrors—all of which are repulsive to the majority—so there are some people who experience delicious shudderings—" des frissons exquis "—when they see a man or an animal in torture or read a description of such things. In the eighteenth century it was

K 2

not unusual for a country cousin on a visit to London to be taken as a treat to see half a dozen men and boys hanged at Newgate, and then to complete the happy day by a visit to Bedlam to see the madmen flogged! Fortunately, public opinion and education seem to have been able actually to alter the operation of the emotions excited by these brutalities—so that to-day practically everyone in the Western States of Europe regards the unnecessary infliction of pain with horror and indignation, and is anxious to avoid witnessing pain, even in cases where it is a necessary evil.

It is a mistake to suppose that there is any tendency on the part of scientific men or medical men to be callous or indifferent to the infliction of pain. The surgeon sometimes has to inflict pain in order to prevent greater future pain or death—but he is not indifferent to the pain he causes. He is not even " cruel only to be kind " —but appears cruel to the unthinking because he has to give pain which he knows will save his patient from far greater pain, and he has to maintain a calm and determined attitude in order to help those around him to exercise self-control. The medical art is, above all things, an art of removing and abolishing pain, and its practitioners are all the more sensitive concerning pain because they know more and see more of it than other people, and make it their chief business to alleviate suffering.

Charles Darwin took a prominent part twenty-five years ago in urging the Government of the day not to make a law which would prevent physiologists and medical men from obtaining knowledge as to animal life and disease by experiment. The great naturalist was a great lover of animals and a most gentle and tender-hearted man. He wrote to me in 1870 : " Experiment must, of course, be allowed for the progress of physiology and medicine, but not for damnable and detestable curiosity. I will write no more about it, or

I shall not sleep to-night." Mr. Darwin was alluding to horrible so-called "experiments" which in former days—especially in the latter part of the eighteenth century—were made by utterly irresponsible and ignorant amateurs, witnessed by fashionable ladies, and reported in the newspapers and letters of the day. It is these reckless and useless "experiments" which rightly excited horror and opposition a century ago, and were described by the name "vivisection." We have to thank these blundering philosophers of the salons of a past age for the mistaken feeling with which some people regard the really valuable and careful investigations which are made by medical men at the present day, with the use of every precaution to prevent pain to the animals used.

The testing of drugs, the inoculation of parasitic disease, and the trial of different modes of removing or controlling the disease so inoculated, carried on by highly trained and learned men, who thoroughly know what they are about, and who communicate with one another from all parts of the world as to the progress they are making in curing or even abolishing diseases, such as diphtheria, cholera, sleeping sickness, and phthisis are very different from the impudent unscientific "experiments" of the days of Horace Walpole. The inquiries carried on in the modern laboratories of our great universities should not for a moment be confused with the horrors performed to glorify and show the superior cold-bloodedness of drawing-room pretenders to science, in those strange times.

I believe that most sensible people feel as Mr. Darwin felt, and I myself would certainly subscribe to what he wrote to me in the letter which I have quoted above. Amongst those who feel thus strongly on the subject there are some who can control their emotion and calmly consider whether the pain inflicted under any given circumstances is justifiable as leading to a great ultimate diminution of pain by the knowledge obtained. There are others who are constitutionally incapable of controlling their emotion in this matter. They hear dreadful stories of cruelty, and are so upset that they are incapable of ascertaining whether the stories are true or not. They are quite unfit to weigh the question as to whether the pain given in the case they hear of may or may not be a necessary step towards avoiding far greater pain in the future for thousands of human beings and sentient animals. Far be it from me to think harshly of these tender-hearted people, though their mistaken outcry may tend to stop the discovery of pain-saving and life-saving knowledge. I feel more sympathy with them than with those (happily rare) individuals who are really indifferent to seeing or giving bodily pain to men or to animals.

There is reason to hope that careful and well-considered statement of the facts will eventually enable many of those who are mentally unhinged by descriptions of pain and bloodshed to recognise that they have been deceived, partly by their own fancies and partly by the false statements of professional agitators. Unfortunately, there are always present in human society individuals who find it to their advantage to excite the minds of their more emotional fellow-citizens by tales of horror. The lust of such power—the power to lead or urge a large body of men driven by emotional excitement into violent action—has led from time to time to exaggeration, misrepresentation, and elaborate plot and perjury directed against a group of innocent or worthy people, whose proceedings were mysterious or misunderstood by the community at large. Thus, from time to time, the crowd has been infuriated and led to the murder of the Jews by agitators, who started the baseless story that the Jews had slain a Christian child, and used its blood at their feast of the Passover. Titus Oates and Lord George Gordon made use of the unreasoning emotion of the crowd in the same way. To a less serious extent the emotional unreasonableness of a number of men and women is being played upon at the present day by quite a large variety of agitators, would-be leaders of crusades and campaigns against the beneficent work of the physiological and medical laboratories of our universities and medical schools.

There are one or two other features about " cruelty " and the mental conditions leading to and arising from it, which, however uncanny and troubling, should be carefully considered when public opinion is roused in regard to its repression. Among these is the fact that the word is freely applied to the mere infliction of pain without consideration of the question as to whether there is a guilty mind determining it. Storms and frosts are called " cruel " by poetic license ; but it is probably quite wrong to call a cat or a tiger cruel. These animals take pleasure in playing with their prey, as they would with an inanimate ball or mechanical toy. There is no reason to suppose that they are conscious of the infliction of pain or take pleasure in pain as pain. And so it must happen sometimes with thoughtless human beings who disregard the pain which they cause, when eagerly engaged in " sport " or in the pursuit of some all-absorbing and consuming purpose. The whole subject of cruelty is a distressing one, but should not on that account be misapprehended or dealt with wildly and blindly.

Twenty-five years ago a Royal Commission sat which was appointed to inquire as to what restrictions, if any, it was desirable to place upon the practice of making experiments on animals for physiological and medical purposes. As a result of its labours an Act of Parliament was passed which made definite regulations for the purpose of preventing unqualified persons from indulging in reckless experiments on animals. There were stories circulated by the agitators then—as there are now—to the effect that medical students perform horrible and painful operations (vivisections, as the agitators term them,) on live animals in secret or with the connivance of their teachers. It was proved twenty-five years ago that these stories were false. At the same time an elaborate law was passed to satisfy the emotional persons misled by the agitators, which made it necessary for an experimenter (1) to have a licence (dependent on a certificate as to his competency); (2) that he should use anæsthetics; and (3) that experiments should only be carried out in licensed laboratories.

The agitators of the present day have by heart-rending stories, similar to those told twenty-five years ago, produced a similar excitement and a similar result, namely, a Royal Commission on Vivisection, which has been occupied for a year and a half in listening to the statements and delusions of those who declare that the law made twenty-five years ago is insufficient, and that all sorts of cruelties are committed by the physiologists and doctors. The Commission has also questioned the leading physiologists and medical men in the country, and listened to their voluntary statements. I have seen the very voluminous report of the evidence thus given on both sides. The various accusations made against the medical men in the conduct of their laboratories have been carefully gone into. It is contended, on their side, that these charges are based on misunderstanding —the misunderstanding which one would expect from an ignorant person with a strong feeling or prejudice in the direction of the misunderstanding. For instance, the fact that chloroform is administered and the animal rendered insensible when operated on, has been overlooked in some of the accounts which excited the so-called " antivivisectors "—notably in the misleading account of " the brown dog." The whole of the evidence should be read by those who are really in doubt on the matter. Probably it will not be long before the Commission reports, and its conclusions will command the very greatest respect, not only because its members include eminent lawyers, medical men and independent representatives who were ready to give an impartial mind to the inquiry, but also because it is obvious that the very greatest care has been taken to obtain the fullest evidence from both sides.

Sir James Fletcher Moulton, one of the Lords Justices of the Court of Appeal, has made a statement to the Commission in defence of scientific experiment which is a masterpiece of persuasive reasoning and lucid exposition. It is somewhat remarkable that there have been and are persons in high judicial office who have shown active hostility to the cause of science and knowledge in this matter owing to their want of acquaintance with the facts and their readiness to be carried away by blind emotion. Lord Justice Moulton, on the other hand, is a scientific man by education and early training, and has come forward to state in a plain and reasonable way what is the view of the matter which commends itself to him. There is reason to hope that his view will be approved by those who read what he says calmly and without bias. His chief point is that many people are willing to admit that it is right to destroy animals (even by methods which inflict great pain on them) when an immediate result of a good and useful kind is to be obtained—as when we kill animals to serve as food or in order to prevent them from injuring us or destroying our crops and stores. Yet these same persons, he points out, by some defect of imagination are unable to see that the gaining of pain-saving or disease-preventing knowledge as the result of inflicting pain and death on a small number of animals justifies us in permitting that pain and death. They are unable to admit the justification because the knowledge and its practical application does not directly and at once follow upon the first commencement of the search for it, and they have not sufficient acquaintance with the matter to enable them to realise and confidently believe that the beneficent result will ensue. The knowledge has to be built up step by step, and the infliction of pain on the animals is separated by an appreciable lapse of time from the beneficent result—which is none the less the result which was aimed at, and the true consequence of the pain inflicted. Putting aside for the moment the fact that in these inquiries the pain is reduced to a minimum by the use of anæsthetics, it would seem that we ought to be able to recognise that the causing of a certain amount of pain to many hundreds of rabbits, and even dogs, is justified by the consequent removal of a far greater amount of pain from thousands of men and animals who are saved from suffering at a later date by the knowledge so gained.

Lord Justice Moulton suggests two cases of the infliction of pain on animals for comparison. Suppose, he says, a ship to arrive in port which (as might easily happen to-day) is infested by plague-stricken rats; there are, perhaps, ten or twenty thousand rats on board. If the rats escaped and landed they might (not certainly, but probably) infect a whole city, even a much larger area, with plague, and cause death and disaster to thousands of human beings. Everyone will agree that the owner of the ship would be justified in destroying all the rats on the ship by sulphur fumes, or whatever other painful method might be necessary to prevent even one from escaping. A vast amount of suffering would be inflicted on the rats in order to prevent a far greater contingent amount of suffering. Now suppose that a man, by infecting some hundreds of rats and other animals with plague, and by trying various experiments on these animals with curative drugs, and by other operations upon them, can in all probability arrive at such a knowledge of plague and how to check it as to enable us to arrest its propagation, and so to save thousands, or even millions, of human beings from this painful and deadly disease, are we to say that this investigator must not carry on his studies, must not find out how to stop plague in future because to do so he will have to give some amount of pain to a hundred or more animals? Clearly, if we justify the shipowner we must justify the inquiries and experiments of the medical discoverer. In both cases we must hold—every sane man really does hold—that it is right to inflict pain with the expectation (not a certainty in either case, but only a reasonable probability) of preventing a far larger and more serious amount of pain in the future. It is the choice of the lesser of two evils.

And thus we are led to admit that it is right that experiments and studies attended with some pain to animals should be carried on, on condition that competent and serious persons make them, for the purpose of gaining increased knowledge of the processes of life and disease. Such studies have already yielded great results —the pain in the wards of hospitals and in sick rooms is not a tenth of what it was a hundred years ago. The death-rate of great cities is a third less than it was fifty years ago. Modern medicine and modern surgery are really and demonstrably immense agencies for preventing pain and the anguish and misery which is caused by untimely death.

A Society for the Defence of Research has been established this year (1908) with the Earl of Cromer as its president. The Society has issued some valuable pamphlets showing what improvements in medical knowledge have been recently effected by means of inoculations and other experiments in which animals have been used though subjected to as little pain and discomfort as consistent with the enquiries made. Ignorant opponents of medical research assert that the scientific study of the processes of life and disease in laboratories has not helped in the great progress in medical practice which marks the last fifty years. But the medical men who are the leaders of their profession unanimously assert, and prove by detailed accounts of the discoveries made, that such study has been essential to the progress established, and is essential for further progress. Lord Lister, who by his antiseptic method of treating surgical wounds has saved more pain to present and future generations of men than all the torturers of the Inquisition ever inflicted or dreamed of inflicting, has been the leader in declaring the inestimable value to humanity—in fact, the absolute necessity—of physiological experiments on animals. Whose judgment on this question can be considered of greater value than his?

The anti-vivisection agitators, for the purpose of exciting the emotions of those who listen to them, use the word "torture" as describing the action of such men as Pasteur and Lord Lister. To torture is to inflict an ever-increasing amount of pain, with the view of "extorting" a submission, a confession, or treasure from a victim. To suggest that scientific and medical men apply pain in this way, and to spread the word "torture" among the ignorant, emotional public, in connection with their inquiries, is dishonest as well as ungrateful.

One valuable result of the work of the present Royal Commission on what is called "Vivisection," but should be called "the use of animals in the discovery of means of controlling disease and alleviating pain," is that it is made quite clear that there is very little pain at all inflicted in this beneficent work, owing to the fact that anæsthetics and narcotics are administered to the animals when anything which might cause pain is done. I do not hesitate to say that there is in this country less pain caused in a whole year in all the laboratories where this great work for the public good is carried on than in a single day's rabbit-shooting.

It is important to correct, if possible, the misunder-

standing which very naturally exists as to what physiologists and doctors mean by "experiment." In ordinary
language an "experiment" suggests a haphazard venture,
the doing of something blindly and in ignorance, just
"to see what will happen." It is true that long ago in
the eighteenth century there were men callous enough
and ignorant enough to make such "fool's experiments"
on living animals. But when scientific men speak of
"the experimental method" and the acquisition of
knowledge by experiment, they do not allude to
haphazard attempts to see what will happen when
something extraordinary is done. The experiment of
the experimental method is arranged so as to provide a
definite answer to a definite question, and the question
has been thought out by a man who knows the whole
record of previous experiment and knowledge in regard
to the subject which is under investigation.

Thus in the inquiry as to the possible prevention of
the deadly effect of snake poison introduced into the
human body by the bite of snakes, the first question
asked was, "Is it true, as sometimes stated, that a
poisonous snake is not poisoned by having its own
poison injected into its flesh?" The experiment was
tried. The answer was, "It is true." Next it was asked,
"Is this due to the action of very small doses of the
poison which pass constantly from the poison gland into
the snake's blood, and so render the snake 'immune,' as
happens in the case of other poisons?" The experiment
was tried. Snakes without poison glands were found to
be killed by the introduction of snake's poison in a full
dose into their blood. Then it was found that a horse
could be injected with a dose of snake poison, or half
the quantity necessary to cause death, and that it
recovered in a few days. The question was now put,
"Is the horse so treated rendered immune to snake
poison, as the snake is which receives small doses of
poison into its blood from its own poison gland?"

Accordingly the experiment was made. The horse was given a full dose of snake poison, and did not suffer any inconvenience. At intervals of two days it was given increasing injections of snake poison without suffering in any way, until at last an injection in one dose of thirty times the deadly quantity of snake poison—that is, enough to kill thirty unprepared horses—was made into the same horse, and it did not show the smallest inconvenience. The question was thus answered: Immunity to snake-bite can be conferred by the absorption of small quantities (non-lethal doses) of snake poison. The next question was this: "If something has been formed in the horse's blood by this process, which is an antidote to snake poison, should it not be possible, by removing some of the horse's blood and injecting a small quantity of it into a smaller animal, to protect that animal from snake bite?" The experiment was accordingly made. Rabbits and dogs received injections of the blood of the immune horse. An hour after they received full doses of snake poison. They suffered no inconvenience at all; they were "protected," or "rendered immune." The next question was, "Will the antidote act on an animal after it has already been bitten by a snake?" The experiment was made. Rabbits were injected with snake poison. After a quarter of an hour they were on the point of death. A dose of the immune horse's blood was now injected into each—in ten minutes they had completely recovered and were feeding. The means was thus found of preventing death from snake-bite. The protective horse-blood was properly prepared, and sent out at once to Cochin China and to India. It was there tried upon human beings who had been accidentally bitten by deadly snakes, and it proved absolutely effective; it saved the men's lives. It is now used (wherever it can be obtained in time) as the sure antidote to snake-bite, though it is not at present possible to supply it whenever and wherever it is needed. That is an example, briefly told, of the experimental questioning of Nature—such as is pursued in the laboratories of medical men and physiologists. They do not perform haphazard experiments; but each experiment is so arranged as to give a definite answer to a definite question, leading to a large result. By no other process can knowledge of many things, which it is urgent for us to have, be obtained. We should have to wait centuries if we merely watched Nature, and hoped for some accidental circumstance to reveal the facts.

What, after all, do we understand and mean by "pain"? It is not merely the sharp sting, and consequent shrinking caused by wounds and violence. That, we know well enough, is a beneficent arrangement by which men as well as animals are prevented from knocking themselves to pieces, and are driven into avoiding danger to life and limb. But "pain" includes, besides this, the anguish arising from the weary, fruitless struggle against disease and starvation, from the disaster to the household caused by the untimely death of its mainstay, from the slaughter of children by poisonous foods, and from the neglect of the laws of health of body and mind.

Ignorance, the "curse of Hell," is the cause of all suffering. Knowledge is the wing which takes us heavenward, and frees us from misery. I cannot put it better than in Shakespeare's words. It is man's destiny to diminish pain on this earth, and that not by timidly shrinking from and emotionally raving about the horrors of pain, but by facing them and deliberately accepting the responsibility of producing a small and brief suffering to a few animals as the price of the salvation of his fellow-creatures from the far greater pain which is the assured and fatal companion of ignorance—accursed ignorance!

A recent writer has told us that he cannot believe

that good will follow from the wilful destruction by man of Nature's greatest and most beautiful production —a living thing. He poses as a sentimentalist and seems to regard it as the indication of a superior and gentle mind to refuse to sanction the removal or even the temporary discomfort of what Nature has called into life. I, too, claim to be a sentimentalist, but the sentiment which thrills me is one of revolt against the needless and remediable suffering of all humanity— suffering which man has brought on himself by his stumbling, half-hearted resistance to Nature's drastic method of purifying and strengthening the race, her remorseless slaughter of the unfit. It is this suffering which some would allow their fellow-men still to endure, now and for generations to come, rather than have their own tranquillity disturbed by the record of that modicum of immediate pain and sacrifice of animal life which is the price of freedom for mankind from far greater pain hereafter. We have to learn to mitigate and to minimise pain, not to run away from it. It is childish to weep over the distortion and destruction of Nature's products by man's violence and ignorance. What we can and should do is to see that our dealings with this fair earth and its living freight are guided not by vain regret, but by knowledge and foresight.

THE END

R. OLAY AND SONS, LTD., BREAD ST. HILL, E.C., AND BUNGAY, SUFFOLK.

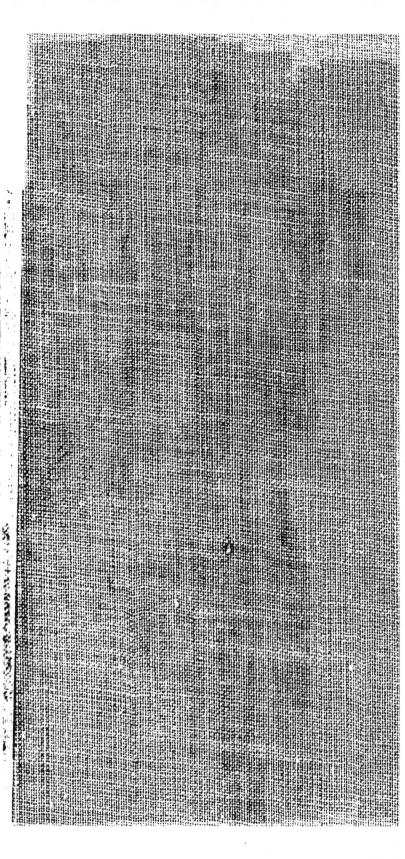